Camel Cigarette Collectibles:

1964-1995

Douglas Congdon-Martin

Schiffer Publishing Ltd

77 Lower Valley Road, Atglen, PA 19310

To Kay Johnson, my aunt, for years of support and love. Thanks.

Designed by "Sue"

ISBN: 0-7643-0196-9
Printed in Hong Kong

Library of Congress Cataloging-in-Publication Data

Congdon-Martin, Douglas.
    Camel cigerette collectibles, 1964-1995/Douglas
Congdon-Martin.
      p. cm.
    Includes bibliographical references (p.).
    ISBN 0-7643-0196-9 (paper)
    1. Smoking paraphernalia--Collectors and collecting.
2. R.J. Reynolds Tobacco Company--Collectibles.
I. Title.
TS2280.C66 1997
688'.4'075--dc21
                       96-30074
                       CIP

Published by Schiffer Publishing Ltd.
77 Lower Valley Road
Atglen, PA 19310
Phone: (610) 593-1777; Fax: (610) 593-2002
E-mail: schifferbk@aol.com
Please contact us for a free catalog.
This book may be purchased from the publisher.
Please include $2.95 for shipping.
Try your bookstore first.

We are interested in hearing from authors
with book ideas on related subjects.

Contents

Acknowledgments

In the Acknowledgments to my book about the early years of Camel Cigarette Collectibles, I thanked a number of people, most of whom also contributed their support to this volume. Odell Farley contributed generously from his vast collection, and provided information, support, and friendship at every step of the process. His friend, Judy Berrier, also provided invaluable knowledge and assistance. Thomas A. Gray, was always there for me, and continues as a friend even after I invaded his home with my array of equipment, taking over his living room. Others supported my efforts in every imaginable way. I hope they are all pleased with the results of their kindnesses. I thank them all. Jack Pennington lent his expertise in assigning current market values to the items represented here.

Assisting me in gathering the images were Dawn Stoltzfus, of Schiffer Publishing, and my daughter, Sarah Congdon-Martin. They both put up with an intense work schedule and a general lack of sleep. Sue Taylor has lent her creativity to the design of this book.

Chapter 1
Camels and the Shifting Sands of Cigarettes

When Richard Joshua Reynolds introduced Camel Cigarettes in 1913, he gave the brand every opportunity to succeed, creating one of the most creative and effective national campaigns to date. Built on the success of his Prince Albert brand of smoking tobacco, Camels immediately won acceptance around the nation. In December, 1913, a very short time after its introduction, Camel was declared to be the "standard brand wherever it has been offered to the public," by the editor of *Tobacco* (Nannie M. Tilley, *The R.J. Reynolds Tobacco Company,* 1985, page 213.)

Before the introduction of Camel Cigarettes the strength of the American tobacco industry was in chewing tobaccos. With Camels, Reynolds changed the course of the industry. The blend was Turkish and domestic burley and bright leaf tobaccos, with Maryland tobacco added in 1916 for its exceptional burning abilities. R.J. himself was in on developing the blend, as were James Walter Glenn, James B. Dyer, and W.N. Reynolds. The blend offered a pleasing taste and aroma that took Camels and R.J. Reynolds to the forefront of the tobacco world.

In 1915 Camel was number one in the U.S. cigarette market, with a 12.55% share. By 1916 the share had doubled to 25.79% firmly ensconcing Camel in the marketplace. They were to hold the lead until 1930, going as high as a 45.07% share in 1923. The upstart Lucky Strike Cigarettes of the American Tobacco Company were introduced in 1925 and overtook first place for the years 1930-1933. In 1934 Camels regained the lead and held it through 1941. In part based on their "Lucky Strike green has gone to war" campaign, an ingenious ploy to capitalize on the patriotic fervor of the time while making a long-delayed shift to white packaging, Lucky Strikes were in first place again from 1941-1948. Camel fought back and recaptured headed the race from 1949-1959, when Pall Mall, one of the new king-sized cigarettes, took over.

1959 was to be the last year of Camel's dominance. In part this was due to the proliferation of new brands on the market. Until 1925, when Lucky Strikes and Chesterfields were introduced, Camel had little real competition. Pall Mall entered the fray in 1936, and had a slow growth until its brief dominance from 1960-1965. This was to be last time a non-filtered cigarette was to lead the industry.

In December, 1952, *Readers Digest* republished an article from the *Christian Science Monitor* entitled "Cancer by the Carton." It triggered a rush toward filter cigarettes that would dominate the 1950s and 1960s. R.J. Reynolds was at the head of the industry again, thanks in part to a 1951 European vacation by Edward A. Darr, vice president in charge of sales. According to Tilley, he learned that at least half of the cigarette market in Switzerland was given over to filter cigarettes. While it was not particularly well-received in Winston-Salem on his return, the idea resurfaced, backed by the power of the RJR President, which Darr became in November, 1952. At his urging the research and development was undertaken. So it was that in early 1954, four months after Dr. Ernest L. Wynder of the Sloan-Kettering Cancer Institute, announced that he had induced cancer in mice by painting them with tobacco tar, R.J. Reynolds introduced Winston, a filter cigarette.

Between 1958 and 1966 Camel and Winston virtually switched places in market share standings. In 1958 Camel held 14.53% of the market while Winston held 9.61%. In 1966 Winston held 14.31%, making it the number one cigarette brand in the nation. Camel had 9.25% of the market.

While always holding an important place of honor in the host of cigarette brands, Camel would never again reach its former heights. Its sister brand, Winston, held the lead from 1966-1974, before being beaten by Marlboro. From 1975 through 1995 Marlboro has led all challengers, a record reminiscent of Camel's earlier days.

Still the name of Camel is nearly synonymous with cigarettes. It has become part of popular culture in a way few other brand names have. And all signs are that it will continue strong into the future.

For real smoking enjoyment

Camel

CAMEL
TURKISH & DOMESTIC BLEND CIGARETTES

DECEMBER 1965

SUN.	MON.	TUE.	WED.	THU.	FRI.	SAT.
			1	2	3	4
5	6	7	8	9	10	11
12	13	14	15	16	17	18
19	20	21	22	23	24	25
26	27	28	29	30	31	

NOVEMBER 1965

SUN	MON	TUE	WED	THU	FRI	SAT
	1	2	3	4	5	6
7	8	9	10	11	12	13
14	15	16	17	18	19	20
21	22	23	24	25	26	27
28	29	30				

JANUARY 1966

SUN	MON	TUE	WED	THU	FRI	SAT
						1
2	3	4	5	6	7	8
9	10	11	12	13	14	15
16	17	18	19	20	21	22
23 30	24 31	25	26	27	28	29

Chapter 2
The Filter Camel: 1964-1973

After the debacle it experienced when it gave the Camel pack a "facelift" in 1957, RJR was reluctant to tamper with its leading brand name. The advertising of the early years of the 1960s is virtually indistinguishable from that of the 1950s. With its fifty years behind it, Camel was a comfortable brand with a loyal consumer base, and there seemed little reason for innovation.

The meteoric rise Winston and the other filter cigarettes, while Camel sales slowed and began to drop, caused a reassessment of all this. If Camel was going to continue to hold a viable position in the marketplace, it would need to become more aggressive. Finally, in 1965 the Camel Filter was introduced, an 85mm cigarette in a soft pack. The idea of a filter-tipped Camel had been around since 1954, when the company went so far as to order the equipment, supplies, and even advertising for such a venture, but the plan was scrapped before production began. (Tilley, p. 498). Even in 1965, the decision was made very carefully. Three basic pack designs were developed, along with advertising for each. They included a brown & white version, a gold foil version, and the traditional pack design, elongated to accommodate the new cigarettes. Apparently the choice came down to the traditional pack and the brown and white, because in October, 1965 executives were given a box of four pack designs of the new filter Camels for their input. They included the brown and white version in three variations, and the traditional design. When it was over, the brown and white with the words "Filter Cigarettes" in the red band and "Famous Camel Quality!" at the bottom was chosen.

In 1966 Camel also introduced a menthol filter variety. It did not do well in the marketplace and was discontinued in 1968.

1966 marked another addition to the Camel pack and to all others. Beginning on January 1, 1966 all cigarette packages had to carry the warning "CAUTION: Smoking May Be Hazardous to Your Health." That was amended on November 1, 1970 to read "WARNING: The Surgeon General Has Determined That Cigarette Smoking Is Dangerous to Your Health." This message was added to advertising in 1972, and continued until October 11, 1985, when it was again amended.

In 1971 Camel Talls, a filtered 100 mm cigarette made a brief appearance but was never fully marketed.

The William Esty Company handled Camel advertising from 1933 until June, 1966, when DFS (Dancer-Fitzgerald-Sample, Inc.) took over the account. As one of its final campaigns, DFS introduced a human-like Camel character. As a race car driver in the GT Challenge, this Camel had none of the poise of the Joe which would follow. He was a cartoonish figure, stuck in a car which was clearly too small for him. But he was driving, wearing a helmet, and, of course, smoking a Camel. From this germ of an idea a major campaign would grow. DFS held it until 1974 when the account returned to Esty.

One other innovation of interest to the collectors is the gradual introduction of Camel premiums. With the exception of a few lighters, pack holders, and ashtrays produced during the 1950s, Camel had remained true to its original concept of No Premiums. But in the mid-1960s the idea of making wider use of the internationally familiar pack logo was irresistible. It began appearing not only on ashtrays, but on glassware, playing cards, and clothing. The first steps were taken in one of the most prolific and successful marketing ventures ever.

Opposite page:
"For real smoking enjoyment - Camel." This 1966 calendar uses the same graphics as 1964's. 1965, 20.5" x 13". $75-90.

"For smoking enjoyment - Camel." Metal
sign, 1964. 12" x 32". $75-100.

"Have a real cigarette - Camel." Cardboard point of sale display, #337-A,
1965. 13.5" x 19.75". $50-65.

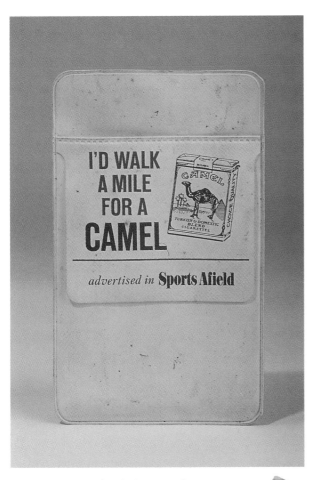

Pocket liner, 1960s. $5-10. *Courtesy of Odell Farley.*

"Camel's real taste satisfies longer." Counter display, #474, 1965. 12" x 9.25" $50-65.

"Ca c'est vivre...Savourer une." Cardboard advertisement in French, c. 1965. This was also made in a paper version. $35-45.

"Fume un verdadero cigarillo." Cardboard advertisement in Spanish, c. 1965. $35-45.

"In any language...America's largest-selling regular size cigarette." 1966 paper calendar in three languages, English, Greek and Spanish, 1965. 16" x 9.5" $50-65

As they prepared to introduce Camel filters in 1965, Reynolds tested three different pack designs, using them in various advertising formats. Three varieties are shown here. Introductory ad for Camel filters, for each of the three pack designs. $35-45 ea.

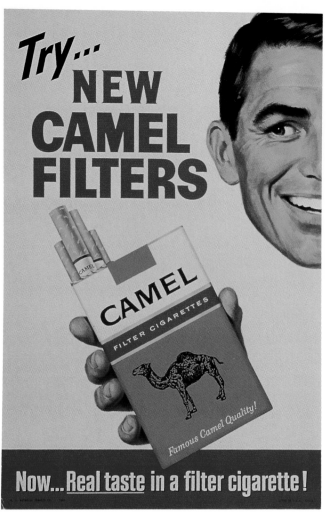

"Try new Camel Filters." Counter display
using the winning pack design, #521B,
1965. 13.75" x 9". $35-45

"Merry Christmas - Give Camel." Paper
sign, #412-CR, 1965. 15" x 28". $50-65.

"Give Camels." Diecut cardboard
Christmas sign, c. 1965. 14" x 14".
$35-45.

Mid-1960s postcard, showing Whitaker
Park. $10-15. *Courtesy of Odell Farley.*

"New King Size, Surprisingly Mild Taste."
Introductory advertisements for King Size
Camels. c. 1965

Ceramic ashtray with large three-dimensional camel in the center, c. 1965. Around the base it reads "Camel Filters." 9.5" x 9.5" $50-65

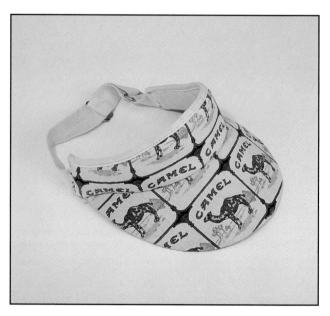

Cotton sun visor, c. 1965. 4" x 11". $5-10. *Courtesy of Odell Farley.*

Camel glass, c. 1965. 4" x 3.5". $25-35. *Courtesy of Odell Farley.*

Plastic Camel mug, c. 1965. 4" x 4". $10-15. *Courtesy of Odell Farley.*

Camel playing cards, c. 1965. 3.5" x 2.5". $50-65.
Courtesy of Odell Farley.

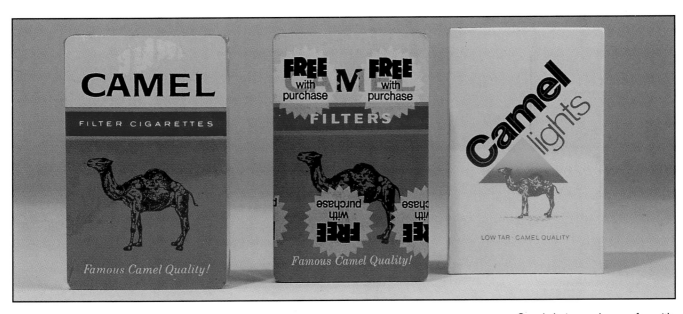

Camel playing cards came free with
purchase. Brown filter packs, 1960s,
lights, early seventies. $10-15 per pack.
Courtesy of Odell Farley.

In any language...

- AMERICA'S LARGEST-SELLING REGULAR SIZE CIGARETTE
- ΤΟ ΣΙΓΑΡΕΤΟ ΚΑΝΩΝΙΚΟΥ ΜΕΓΕΘΟΥΣ ΜΕ ΤΗΝ ΜΕΓΑΛΥΤΕΡΑΝ ΚΑΤΑΝΑΛΩΣΙΝ ΕΙΣ ΤΗΝ ΑΜΕΡΙΚΗΝ
- EL CIGARILLO DE TAMAÑO REGULAR DE MAS VENTA EN AMERICA

NOVEMBER 1966

SUN	MON	TUE	WED	THU	FRI	SAT
		1	2	3	4	5
6	7	8	9	10	11	12
13	14	15	16	17	18	19
20	21	22	23	24	25	26
27	28	29	30			

"In any language America's largest-selling regular size cigarette." Variation of the 1967 paper calendar with English, Greek and Spanish text, 1966. 16" x 9.5". $35-45

America's largest-selling regular size cigarette

NOVEMBER 1966

SUN	MON	TUE	WED	THU	FRI	SAT
		1	2	3	4	5
6	7	8	9	10	11	12
13	14	15	16	17	18	19
20	21	22	23	24	25	26
27	28	29	30			

"America's largest-selling regular size cigarette." Paper 1967 calendar, 1966, also published in Spanish. 16" x 9.5". $35-45.

El cigarrillo de tamaño regular de más venta en América

NOVIEMBRE 1966

"El cigarrillo de tamaño regular de más venta en América." 1966-67 Calendar. 16" x 9.5". $25-35.

"Try New Camel Menthol." Introductory advertising poster. 1966. $50-75.

"Camels real taste satisfies longer!." Paper calendar, 1966-1969. 16" x 9.5" $50-65.

"Try new Camel Menthol." Introductory display rack sign, #578, 1966. 19.25" x 19.5". $75-100.

17

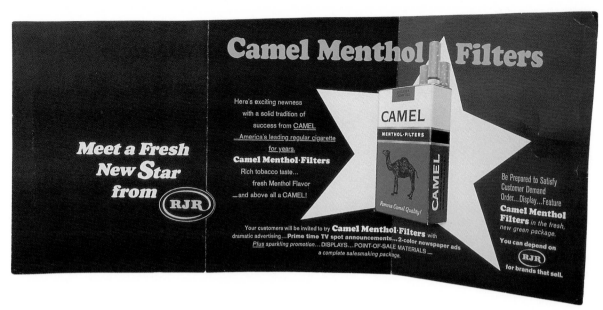

"Camel Menthol Filters." Flyer of instructions to dealers for the use of point of sale advertising materials, 1966. 10.75" x 24.5". $35-45.

"Have a real smoke! Have a Camel Filter!" Camel Filters advertisement, 1967. $20-30.

"Camel Filters satisfy longer without being longer." Camel Filters advertisement, 1968. $20-30.

"Special offer. 2 decks of cards, only 75 cents." Camel Filters cardboard counter display with coupons, #674-2, 1968. 8.75" x 10.25". $25-35.

Christmas carton with decoration at the gold foil end. 1970. 85 mm. $20-30.

Camel tip tray, tin. 6.75" x 5". c. 1970. $20-30. *Courtesy of Odell Farley.*

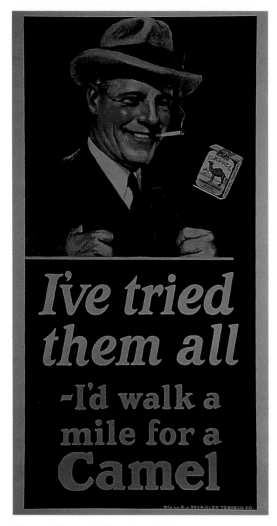

"I've tried them all - I'd walk a mile for a Camel." A poster based on the 1920 original, 1971. 27.5" x 14.25". An identical poster #671-c, was produced in 1973. $125-175.

"Walk Tall." Diecut point of sale advertisement for Camel Talls, 1971. $20-25.

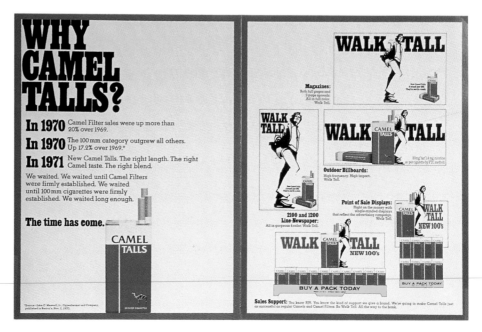

"Why Camel Talls?" An advertisement for the trade introducing Camel Talls and the types of promotion they would receive. 1971. $10-15.

"Walk Tall." Diecut carton insert for point of sale.
1971. $20-25.

"I'd walk a mile for a Camel." Cardboard
poster, #671-c, 1973. 19" x 28". $50-65.

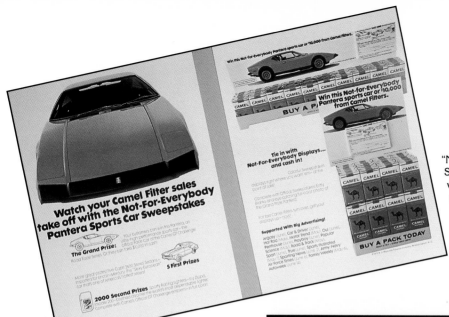

"Not-For-Everybody Pantera Sports Car Sweepstakes." Advertisement introducing vendors to an upcoming promotional campaign, 1973. A prototypical Joe Camel-like cartoon figure is on the Zippo lighter that is offered as second prize. $10-25.

"Free Racing Decal." This advertisement offered a free Camel GT Challenge decal with two packs of filter cigarettes, 1973. Again, the camel is prototypical of the Joe Camel that would be introduced a few years later. $20-30.

Cigarette Pack Art. A book by Chris Mullen and featuring "A unique blend of cigarette pack art." 1979. $35-45.

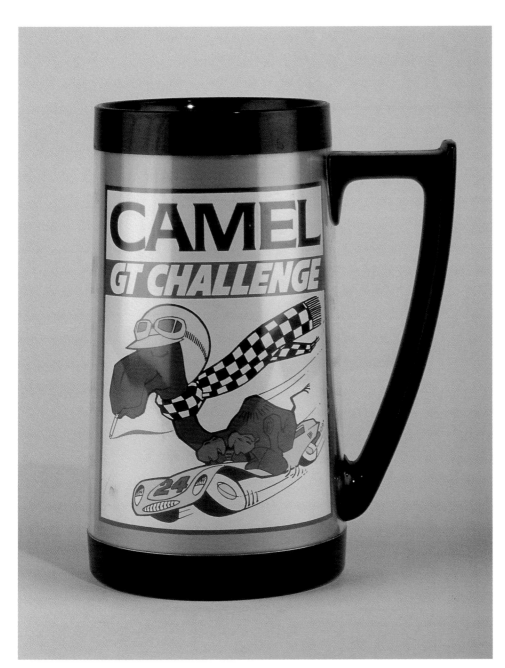

"Camel GT Challenge." Plastic mug with Old Joe riding a race car. This is a precursor to the Joe Camel character introduced in 1990. 6.25" x 5" x 6". $10-20.

The pinback and the pen are from c. 1973 and the lighter is from 1994. While the images of Joe in the lighter and the pinback are miles apart in detail, the idea of creating a human-like camel as a symbol obviously has it roots here, and will develop over the next twenty years. Pinback (c. 1973), and pen. $10 each. *Courtesy of Odell Farley.*

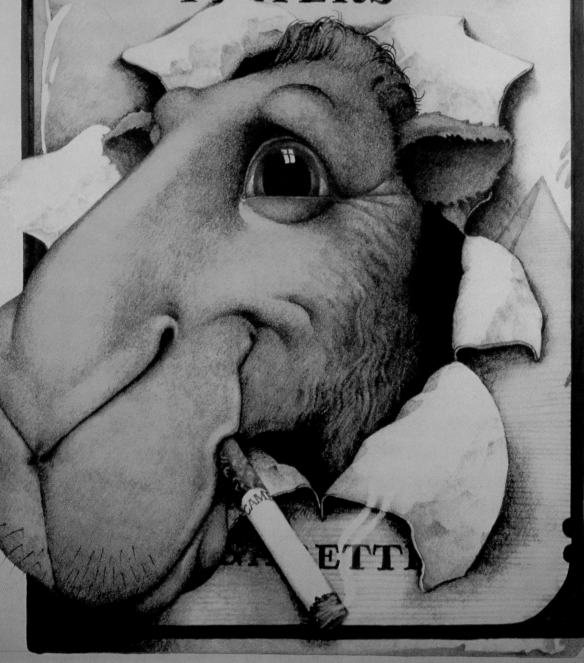

Chapter 3
Vive la Camel: 1974-1987

In 1974 the French subsidiary of R.J. Reynolds asked its agency for an advertisement promoting Camel Filters to run in *Lui*, a French men's magazine. Nicholas Price, an English artist, was commissioned for the design.

The image Price created was of a rather likeable head of a camel bursting through a background very much like the front of a Camel Filters pack. His crooked smile and soulful brown eyes make immediate contact with the observer. As price described it in a 1993 interview, "I gave the camel sort of a twinkle in the eye, and sort of a conspiratorial expression to say, 'I know what you're thinking.'" (*Winston-Salem Journal*, September 2, 1993, p. 23). The advertisement was successful enough to be used in some other European countries. It was also used on some promotional items, but no one expected that it would be at the center of a highly successful and controversial advertising campaign some 14 years later.

In 1975 R.J. Reynolds celebrated its 100th Anniversary. They issued a wonderful book celebrating the company's history and its present success, *Our 100th Anniversary: 1875-1975*. Recounting the successes and some of the rare failures, pride is taken in the fact that "on the threshold of its second century Reynolds remains the industry leader: the only tobacco company to have three cigarettes in the top 10 brands, the top-selling little cigar, the Number 1 smoking tobacco, and the leading plug chewing tobacco." (p. 23) It also recounts its international successes beginning in Europe and continuing in Asia, Canada and South America. In 1974 R.J. Reynolds acquired Macdonald Tobacco Inc., Canada's largest tobacco manufacturers. By 1975 the total international cigarette consumption was estimated to 2,000 billion annually, four times the number smoked in the United States. With sales in 140 countries it was a market Reynolds planned "to make the most of."

Finally, the anniversary book was a celebration of Reynolds diversification into other areas of business. R.J. Reynolds Industries was created in April, 1970. It oversaw a company involved not only in tobacco, but in aluminum, international shipping, foods, and petroleum. In 1975 Reynolds Industries had sales and revenues of well over $4 billion and 33,600 employees.

During all of this Camel remained a cultural icon. The familiar pack and the curved Camel logo began popping up everywhere: lighters, keychains, bath towels. The age of advertising through pasting your name on everything imaginable was dawning, and the Camel name was among the most highly recognized and used. Tom Robbins even made the pack of Camels a central character in his novel *Still Life with Woodpeckers* in 1980.

A trend which began in the early 1970s continues to grow through the 1980s. With the cessation of television and radio advertising, Camel and other cigarette companies began to invest in sporting events. Races like the Camel Pro or the Camel Trophy '96 got the name before the public without actually advertising the product.

The William Esty company resumed its management of Camel advertising in 1974 and continued through 1979, when the account went to BBDO (Batten, Barton, Durstine & Osborn, Inc.). BBDO had the account from January, 1980 until July, 1983, when McCann-Erickson took over. The 75th Birthday campaign was a special project developed by Trone Advertising, Inc., who introduced the figure of Joe Camel for the event. Though Joe was original conceived for use only in the Birthday campaign, he was such a hit that McCann-Erickson continued him in their work, developing him further, including his blue eyes and "smooth character."

Opposite page:
The prototype for the latest Joe Camel originated in this French advertisement for Camel Filters, 1974. 23.5" x 15.5". $150-175. *Courtesy of Thomas A. Gray.*

In this period Reynolds continued to broaden the Camel line. Camel Lights in a soft pack were introduced in 1977 and sales soared in 1978. Camel Lights in a 100 mm size, soft pack, were introduced in 1979, followed by Camel Lights in a hard pack in 1980. 1981 and 1982 saw the introduction of Camel Filters in the hard pack, and in 1987 100 mm Camel Filters were introduced in a soft pack.

Two Camel lighters with the French logo. Note the blue eyes on Camel. c. 1974 2.5" x 1.5". $20-30. *Courtesy of Odell Farley.*

Camel lighters, featuring the French logo that was a prototype for Joe Camel. The one on the right is from the French advertising campaign. $20-25 each. *Courtesy of Odell Farley.*

Keychain and refrigerator magnet, c. 1991. $5-7 each. *Courtesy of Odell Farley.*

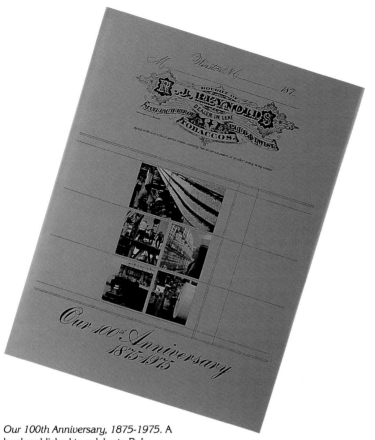

Our 100th Anniversary, 1875-1975. A book published to celebrate R.J. Reynolds' century in the tobacco business, 1975. $20-25.

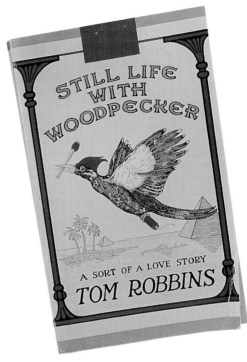

Still Life with Woodpecker. A novel by Tom Robbins which uses the graphics of a Camel cigarette pack in its plot. Copyright 1980, Bantam books. $20-25. *Courtesy of Thomas A. Gray.*

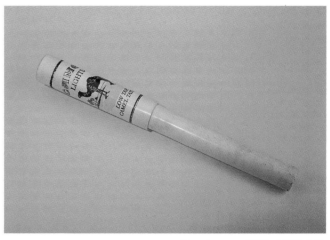

Camel pen, c. 1980s. 5.25". $10-15. *Courtesy of Odell Farley.*

Glass ash tray. 6.75" x 4.5". c. 1980. $10-15. *Courtesy of Odell Farley.*

Camel lighter.
Made in
France. 1980s.
$20-25.
*Courtesy of
Odell Farley.*

Camel Lights frisbee, 1980s. $10-15.
Courtesy of Odell Farley.

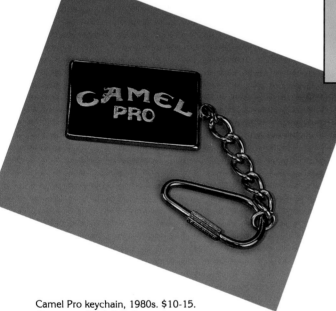

Camel Pro keychain, 1980s. $10-15.
Courtesy of Odell Farley.

Novelty box, used for
packaging John
Henry handkerchiefs,
early 1980s.
Replicans, Bedford,
England. 5.5" x 3.5"
x 2". $20-25.

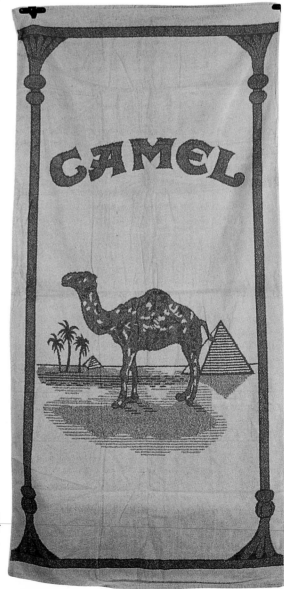

Camel beach towel, 57" x
28.5". $20-25. *Courtesy of
Odell Farley.*

"Everest As It Has Never Been Seen." Poster advertising a program by Ned Gillette and Jan Reynolds who participated in the 1981-1982 Camel Expedition, circling Mount Everest. The program was to be held at the RJR World Headquarters in Winston-Salem, 1982. $10-15.

"Pack Lites." The first Camel Pack Lite cigarette lighter, 1984. $10-15. *Courtesy of Odell Farley.*

Plastic encased 3,238,224,993,405th Camel, to celebrate Tobaccoville Factory Groundbreaking, October 29, 1982. $35-40.

Front and back of a box of Camel matches, 1984. 2" x 1.25". $2-4. *Courtesy of Odell Farley.*

Solid brass Camel GT ashtray, 1984. Made in Taiwan. 3.5" x 6". $12-15. *Courtesy of Odell Farley.*

Left: "Free Pocket Tools with 2-Pack Purchase." A free flashlight and screwdriver on a card with two packs of Camels. Right: Camel pocket protector holding a flashlight, a screwdriver with brush, and a screwdriver, 1985. $10-15. *Courtesy of Odell Farley.*

Camel Pro plastic lighter and ashtray, 1985. $10-15 each. *Courtesy of Odell Farley.*

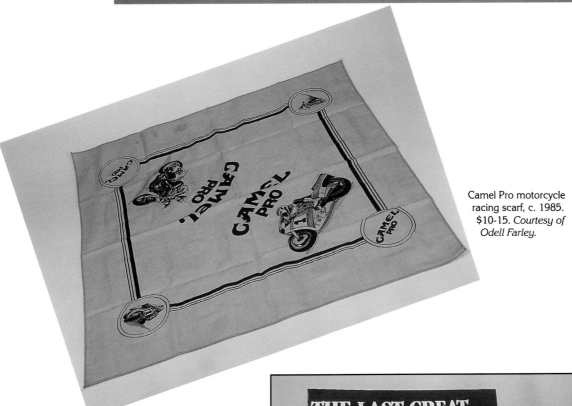

Camel Pro motorcycle racing scarf, c. 1985. $10-15. *Courtesy of Odell Farley.*

Program and decal from Australia Camel Trophy '86. $10-12 each. *Courtesy of Odell Farley.*

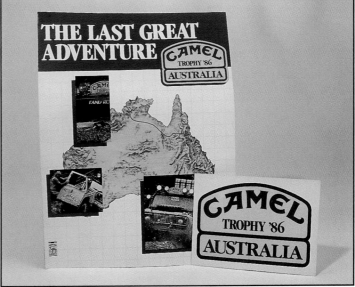

Glass Camel mug, "Summer of '86" and surfboard on back. 6" x 3.5". $10-15. *Courtesy of Odell Farley.*

"Camel. Force." Camel cap with soccer ball and player logo, late 1980s. $20-23. *Courtesy of Odell Farley.*

Brass No. 5, World War II-style trench lighter with leather pouch, late 1980s. $25-30. *Courtesy of Odell Farley.*

Chapter 4
75 And Still Smokin': 1988-1995

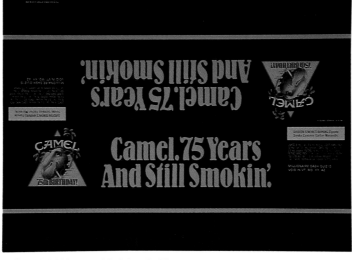

"Camel. 75 Years and Still Smokin'."
Cardboard fold-over sign n, 1987. 15" x 20.75." flat. $20-30.

R.J. Reynolds was enough of a visionary to imagine that his beloved Camels would be around in 1988, 75 years after he conceived them. And he, who approved the famous "Camels are coming!" ads of 1913, doubtless would have loved the excitement that the new campaign created. He may even have anticipated some of the furor that would surround the advertisements. Protests against cigarettes and cigarette advertising were almost as vehement at the turn of the century as they are today. But no one could have foreseen the intensity of the debate over the advertising of cigarettes using an anthropomorphic camel named Joe.

Joe's ancestry, of course, begins with the Old Joe, the camel with the Barnum & Bailey Circus. His image has graced the package of Camel Cigarettes almost from the beginning until today. In the 75 years that followed Old Joe became one of the most recognized symbols in the world. It is an association so strong that it is hard to see a camel without thinking of the cigarette. But despite this strong connection, Old Joe did not carry many of the peripheral associations that advertisers seek. While he has taken on an aura of dignity over the years, it is hard to think of a camel as beautiful or stylish or sexy. He is not sleek or powerful or "cool."

In the early 1970s a couple attempts were made at stretching the archetype of the Old Joe. In 1973 a camel figure showed up driving a race car, helmeted and smoking. While more comic than "cool" it did have a camel doing something other than just standing there, and

something exciting at that.

In 1974 another camel appeared, this time bursting through a pack of Camel Filters in an advertisement in the French magazine *Lui*. All that showed was the head, but the artist had given it an unmistakably human persona.

The 75th Birthday offered the opportunity to create a new image for the old icon of fine cigarettes. Searching for ideas, the French advertisement was dusted off, refurbished, and tinkered with until a new Joe Camel emerged. The principle image of the anniversary campaign has Joe bursting through a pyramid, a few palm trees emerging with him. It is clearly based of the French Joe, but there are a few subtle differences. The French Joe had almost scraggly hair, while the new Joe was neatly coifed. The facial hair and the bristles of a moustache of Joe Francais, would not do for the smooth Joe of 1988. He was clean shaven, a camel in human skin.

Indeed the metamorphosis from dromedary to humanoid was complete. Joe had become bipedal, hands and feet replaced hooves, and there was no trace of a hump. In addition to the image of Joe breaking through the pyramid, the anniversary advertisements showed him engaged in other pursuits. We find him as a well-equipped fisherman in a mountain lake, a race car driver, a Motorcross racer, and partying with his camel friends in the "Wild Pack." We even see him don sunglasses, emblematic of the suavity of the new Joe Camel. As part of the anniversary campaign Joe's image was offered on a variety of merchandise, including a t-shirt, a lighter, and party gear, given away with proof of purchase. Other merchandise would follow.

Joe Camel was an immediate hit, and from 1988 to the present he has been the leading spokes-"man" for Camel cigarettes. His image has been on print ads and billboards, and around him has grown a number of other Camel characters. These include the members of Joe's band, the Hard Pack, Floyd, Eddie, Max, and Bustah, and the Wides Guys, Max and Ray.

Such was the popularity and success of this advertising direction that the demand for Joe Camel merchandise grew. It started with premiums offered with multi-pack cigarette purchases. Then, in 1991, Reynolds introduced a premium program, Camel Cash, which offered a variety of items bearing images of Joe Camel and his friends, as well as traditional Camel logos. Catalogs of merchandise were issued once or twice a year, along a special 80th Anniversary catalog in 1993, a Smokin' Joe's Racing catalog in 1994, and others.

The success of the Joe Camel advertising strategy is credited with shoring up Camel's place in the market, but along the way it has engendered more than its share of controversy. The creator of the French advertisement, Nicholas Price, sued Reynolds for copyright infringement in 1992, claiming that Reynolds had contracted with him only for limited usage of the 1974 design. Reynolds settled with Price out of court. Then, as the Joe Camel image faced criticism from anti-smoking forces for its appeal to children, Price emerged again to express his outrage.

The fact is that Joe has been at the center of a national debate, with one side calling for his ouster and severe limitations on the images used in tobacco advertising, and the other claiming that charges of Joe's influence on children are unsubstantiated. In 1991 an array of health advocates, including the American Cancer Society, the American Lung Association, the American Heart Association and others, petitioned the Federal Trade Commission to ban Joe Camel on the grounds that he appeals to an illegal, underaged market. In June, 1994, the FTC declined, deciding that there was not compelling evidence to support the charges or prove that the ads actually caused young people to smoke. This was an important victory for Reynolds, though it did not quell the criticisms. In a 1995 interview Maura Ellis, a spokeswoman for Reynolds, is quoted as saying, "There is nothing nefarious behind Joe Camel. Joe has been scrutinized up one side and down the other by the U.S. Government and has been exonerated." ("Joe Camel Leads the Pack in Lighting Up Controversy," Sheryl Stolberg, *The Los Angeles Times*, August 21, 1995.

The period of 1988-1995 saw the introduction of several Camel varieties. In 1990 Camel Ultra Lights came on the market in three varieties: 85 mm soft pack, 83 mm hard pack, and 100 mm hard pack. 1991 brought Camel Filters and Camel Lights in 99 mm sizes, hard pack. 1992 was the year of the Camel Wides Filters and Lights, both of which were 79 mm and in hard packs. Camel Special Lights were marketed in 1993 in three varieties: 85 mm soft pack, 83 mm hard pack, and 100 mm hard pack.

Camel's advertising account was with McCann-Erickson from 1983 to October, 1989. From then until October, 1991 it was placed with Young & Rubicam. Mezzina/Brown have had the account since October, 1991.

"Camel. 75 Years And Still Smoking." Materials from the Millionaire cash quiz, published in 1987. The measurements are: 3.25" x 18.25." (top), 3.25" x 5." (bottom left), 4.5" x 9.25." (center right), and 2.5" x 14." (bottom right). $10-15.

"Camel. 75 Años Y Siempre En La Onda!." 75th birthday materials in Spanish, 1987. $5-7 each.

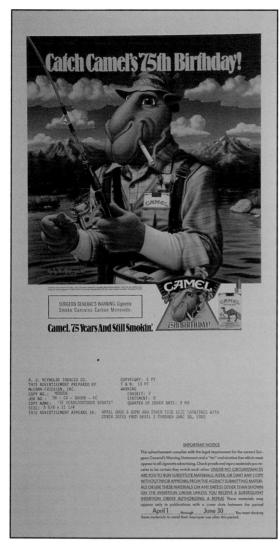

"Free Lighter with 2-Pack Purchase."
Display advertisements in English and
Spanish, 75th birthday, 1988. 10.5" x
13.5". $10-15 each.

"Catch Camel's 75th Birthday!."
Advertising sleeve, 1988. 19" x 10.5".
$20-25.

"Camel. 75 Years and Still
Smokin'." Ad slick for the
75th birthday, 1988.
13.25" x 10.625". $20-
25.

"Camel. 75 Years And Still Smokin'."
Advertising sleeve, 1988. 16.25" x
19". $20-25.

"Fire Up with Camel's 75th Birthday!."
Advertising slick with Joe and his sports
care, 1988. 17" x 13.25". $20-25.

"Gear Up With Camel's 75th Birthday!."
Advertising slick with a motorcycle racing
theme, 1988. 16.5" x 13.25". $20-25

"What's 75 Years Old, Travels in Packs,
and Stands for Great Taste?" Fold out
advertising slick shown closed and open,
it appeared in *Sports Illustrated* and
Penthouse magazines, 1988. 11" x 23.5".
$20-30.

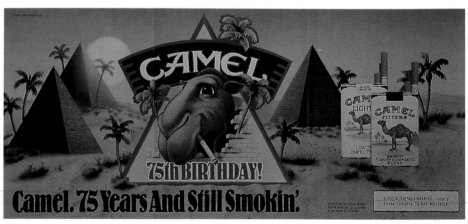

"Party with the Wild Pack." A mailer in which Joe offers coupons for free cigarettes and a T-shirt. $20-25

"Camel 75 Years and Still Smokin'."
Cardboard sign advertising free party gear with a 2-pack purchase, 1988. 18.25" x 19.5". $20-25

"Get On Track with Camel's 75th Birthday!." Advertising slick emphasizing auto racing. Prepared by McCann-Erickson, Inc., for use between April 1 and June 30, 1988. 16" x 12.5". $20-25.

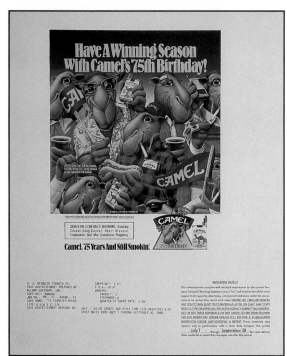

"Have A Winning Season with Camel's 75th Birthday!." Advertising slick for use from April through June, 1988. Prepared by McCann-Erickson, Inc. 16.75" x 14.25". $20-25.

The Great Adventure by Eric Tschum, 1988. This book depicts Camel Trophy adventures from 1981-1988. 10" x 13.75". $20-25.

"Free Party Gear." & "Free Party Mug." 75th birthday advertising flyers with promotional gifts for 2-pack purchases, 1988. Top: 4.5" x 9.25"; bottom: 3.25" x 10.5". $10-15 each.

Mahogany wooden box with an embossed metal emblem commemorating Camel's 75th birthday, 1988. The box contains four glass coasters embossed with the same commemoration. 5" x 10" x 5.5". $25-35.

"Camel. 75th Birthday."
Clothing commemorating
the 75th birthday of the
Camel brand, including a T-
shirt, sweatshirt, and
baseball cap, 1988. Shirts
read "Camel. 75 Years and
Still Smokin'." Shirts: $20-
25. *Courtesy of Odell
Farley.*

Camel "centerfold" from *Playboy,* announcing the 75th birthday, March, 1988. The
figure is basically that of the French
campaign of 1977. $20-25. *Courtesy of
Odell Farley.*

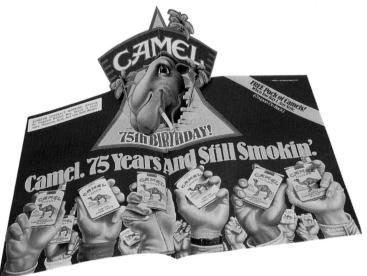

This Camel centerfold announcing the 75th birthday had a chip that enabled it to play music. *Rolling Stone* March 10, 1988. $20-25. *Courtesy of Odell Farley.*

Special Camel 75th anniversary packs and the countertop display introducing them. In his jeans and blue shirt, this Joe is not the smooth character of the years that follow, but the figure is evolving. Counter-top display: 11" x 12". $35-40. *Courtesy of Odell Farley.*

Pop-up Camel 75th birthday announcement, offering free T-shirt. $20-25. *Courtesy of Odell Farley.*

"Camel. 75 Years and Still Smoking." 75th Birthday poster, 1988. 34.75" x 26.5". $50-65. *Courtesy of Odell Farley.*

75th Camel birthday banner, it unfolds to reveal several pennants on a string. $35-40. *Courtesy of Odell Farley.*

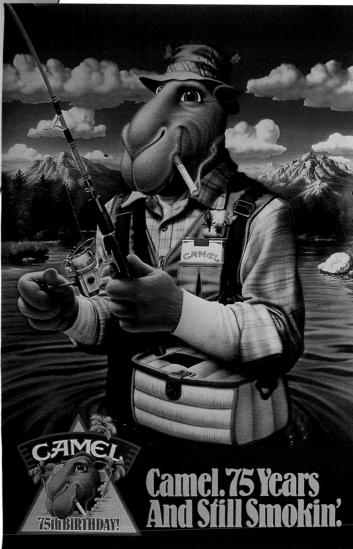

75th Birthday beach towel, 1988. 57" x
28.5". $20-30. *Courtesy of Odell Farley.*

An assortment of promotional items celebrating Camel's
75th birthday including mug, car plate, butane and Zippo
lighters, and pen, all 1988. Mug, $10-12; license plate, $10-
12; pen, $5-7; butane lighters, $10-12; Zippo, $35-40.
Courtesy of Odell Farley.

75th
Birth-
day auto
s h a d e , 1988. 21" x
52.5". $20-30. *Courtesy
of Odell Farley.*

Camel lighters. The lighter on the right commemorates the 75th anniversary, 1988. 2.5" x 1.5". $10-20. *Courtesy of Odell Farley.*

75th Birthday playing cards, napkins, and cloth patch, 1988. Napkin, $5-7; patch, $5-7; cards, $10-13.

Glass tumbler and ashtray, both with the Camel 75th birthday logo, 1988. $5-7 each. *Courtesy of Odell Farley.*

Baccarat crystal camel with cherry base, dated 1988, was an executive gift in honor of Camel's 75th birthday. 8.75" x 5". $500-650. *Courtesy of Thomas A. Gray.*

Detail of medal.

Top, bottom, and inside shots of Camel 75th commemorative birthday tin given to R.J. Reynolds employees on October 13, 1988, with original box. They were signed by Edward A. Horrigan, Jr., C.E.O., and the bottom of tin is inscribed "75 Years of Quality and Tradition, 1913-1988, Special Edition." 3.5" x 3.25". $10-15. *Courtesy of Odell Farley.*

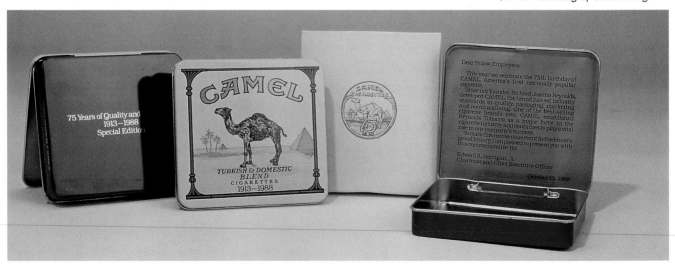

75th Birthday wall clock, 1988.
26" x 13.5". $225-275.
Courtesy of Odell Farley.

Keyring and pin celebrating the 75th
birthday of Camel Cigarettes with
early version of Joe Camel used for
the 1991 United Way Campaign in
Winston-Salem. $5-7 each.
Courtesy of Odell Farley.

"Camel." Plastic Camel
mug, 3.5" x 3.5". c. 1988.
$10-13. *Courtesy of Odell
Farley.*

Lighter holders featuring Joe Camel in regular clothes, tuxedo, and Hawaiian shirt, c. 1988. $12-15 each. *Courtesy of Odell Farley.*

Camel wristwatches. Left to right: 75th Birthday, 1988, $75-100; three hand-painted watches from 1988, $100-125; Smooth Character; Joe Camel; Classic Camel, designed by an employee, $60-75; two wristwatches from Camel Cash Catalog 1; Club Camel from Catalog 2; Joe's Journey Watch, with secondhand is motorcycle Joe, from Catalog 3; Classic Camel Collector's Watch, from Catalog 3; Joe Camel White Tuxedo Watch, from Catalog 4; Classic Camel Onyx Watch from Catalog 5. All watches not priced above: $40-50 each. *Courtesy of Odell Farley.*

"Your free CAMELflage cap has arrived." In 1989, before the advent of Camel Cash, Camel offered this camouflage cap free for wrappers. It came in the camouflage box shown here. The close-up shows that the camouflage included the graphic of Joe Camel. $10-15. *Courtesy of Odell Farley.*

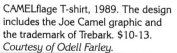

CAMELflage T-shirt, 1989. The design includes the Joe Camel graphic and the trademark of Trebark. $10-13. *Courtesy of Odell Farley.*

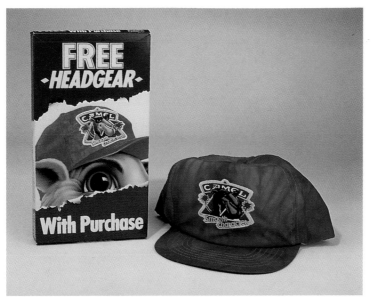

"Free headgear with purchase." Camel baseball cap, free with purchase, 1990. The logo has changed somewhat from the anniversary image. The pyramid is now a yellow triangle with a blue field. Beneath it is written "Smooth Character." Note the blue eyes in the advertisement and the sunglasses in the logo. $20-25. *Courtesy of Odell Farley.*

"Free Camel corduroy cap." Camel baseball cap included with three pack purchase, 1990. $15-20. *Courtesy of Odell Farley.*

"Free 6-pack Tube Cooler." Over-the-shoulder six-pack can cooler, 1990. 29" x 4". $15-20. *Courtesy of Odell Farley.*

49

Camel key card padlock,
1990. 3" x 1.5". $10-12.
Courtesy of Odell Farley.

"Free Camel T-Shirt." Joe Camel Smooth
Character T-shirt, 1990. Free with a three
pack purchase. $20-25. *Courtesy of
Odell Farley.*

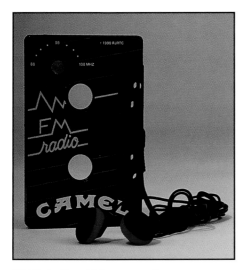

"FM Radio Camel." Premium with purchase, 1990. 4" x 2.5". $10-15. *Courtesy of Odell Farley.*

"A Pack of Camels." T-shirt, 1990. $20-25. *Courtesy of Odell Farley.*

International Camel Motorsport decals, 1990. $10-15. *Courtesy of Odell Farley.*

Camel painter's cap, 1990. $10-12.
Courtesy of Odell Farley.

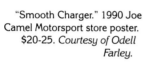

"Smooth Charger." 1990 Joe
Camel Motorsport store poster.
$20-25. *Courtesy of Odell
Farley.*

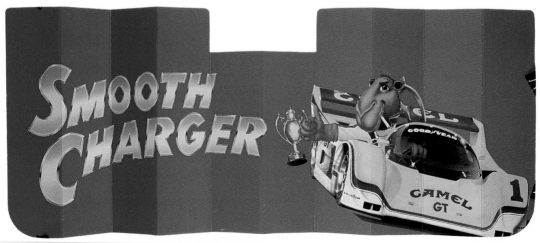

Smooth Charger Camel GT auto shade,
1990. 21" x 52.5". $20-25. *Courtesy of
Odell Farley.*

"Smooth Flyer." 1990
Motorsports store poster.
$20-25. *Courtesy of
Odell Farley.*

Plastic Joe Camel cup,
made for convenience
stores, 1990. 6.5" x 3".
$5-7. *Courtesy of
Odell Farley.*

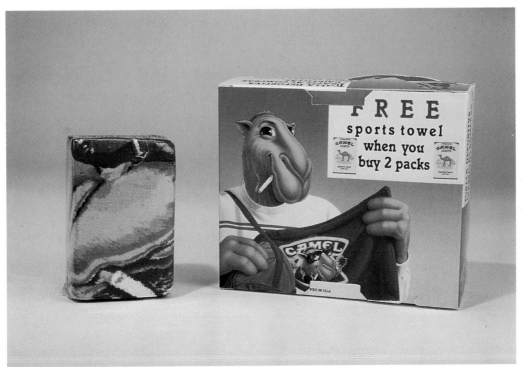

"Free sports towel when
you buy 2 packs."
Sports towel, shown
here in its original
wrapping, came with
two packs of Camels,
1990. The box is written
in English and Spanish.
$10-15. *Courtesy of
Odell Farley.*

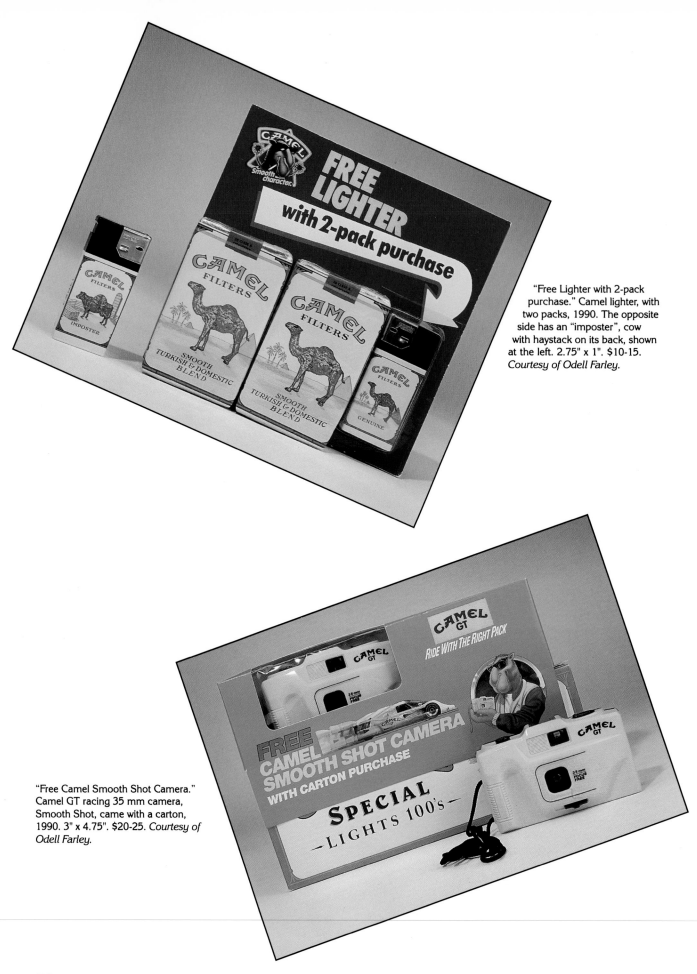

"Free Lighter with 2-pack purchase." Camel lighter, with two packs, 1990. The opposite side has an "imposter", cow with haystack on its back, shown at the left. 2.75" x 1". $10-15. *Courtesy of Odell Farley.*

"Free Camel Smooth Shot Camera." Camel GT racing 35 mm camera, Smooth Shot, came with a carton, 1990. 3" x 4.75". $20-25. *Courtesy of Odell Farley.*

"3 Free Packs when you buy 3." Six pack box, 1990. 3.5" x 5.75". $20-25. *Courtesy of Odell Farley.*

"Smooth Charger" T-shirt with a Camel GT front pocket reads "Camel Mid Ohio Sports Car Course," 1990. $20-25. *Courtesy of Odell Farley.*

"The Premiere of Smooth Magazine." This packet includes a letter and a press copy of Smooth Magazine, c. 1990. While a great idea, the magazine did not stay in print very long. $10-15. *Courtesy of Odell Farley.*

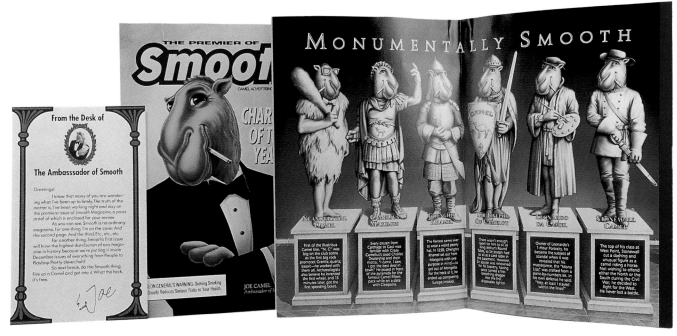

Joe Camel Christmas card, c. 1990. $10-15.
Courtesy of Odell Farley.

"Hollywood-Smooth Character." Pop-up
centerfold magazine advertisement. $10-
15.

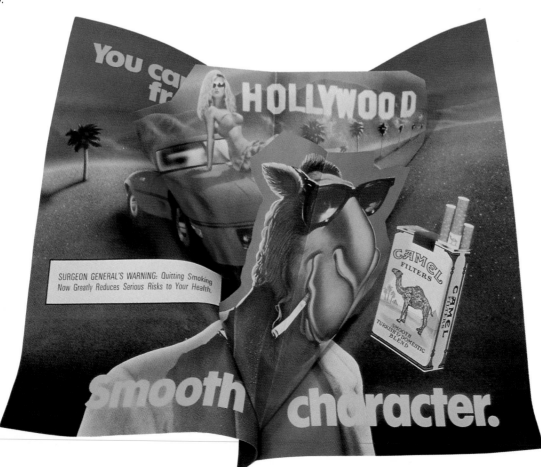

Neon sign, 1990.
$225-275. *Courtesy of
Odell Farley.*

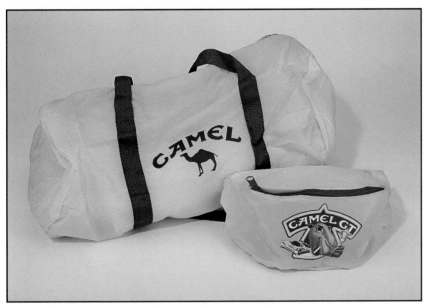

Traditional Camel sports bag and Joe GT
fanny pack, 1990. $10-13 each. *Courtesy of
Odell Farley.*

Four sizes of Smooth Character point-of-
sale signs, one side English and one side
Spanish, 1990. $5-7 each. *Courtesy of
Odell Farley.*

"Smooth Character." Series of 8 Smooth Character posters, 1990.
34.75" x 26.5". $25-30 each. *Courtesy of Odell Farley.*

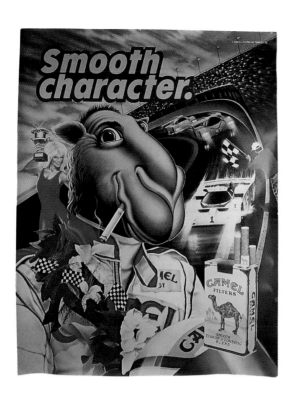

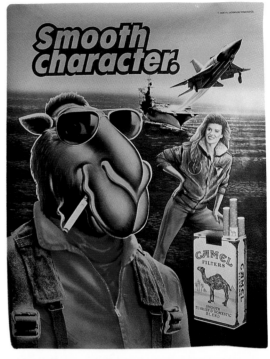

Camel Zippo lighters, 1990s. $35-40 each. *Courtesy of Odell Farley.*

Camel lighters, 1990s. The lighter on the left features Joe Camel, Smooth Character. The center two lighters have a text logo, and the pack design appears on the right lighter. 3" x 1.5". $10-15 each. *Courtesy of Odell Farley.*

Flip-top lighters. Left to right: Camel Light, Special Lights, Joe's Fish & Game Club, and Camel Lights Hard Pack. $10-15. *Courtesy of Odell Farley.*

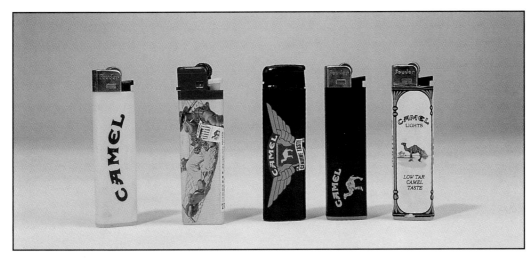

Assorted Camel butane lighters. L to R: Plain logo on white, Joe Camel in hammock made for Japan, Camel Genuine Taste in blue wings, plain logo, Camel lights. 3" x 0.75". $10-13. *Courtesy of Odell Farley.*

Camel ceramic ashtrays. Made in Spain. 3.5" x 8" and 3" x 4". Large: $15-20; Small: $10-13. *Courtesy of Odell Farley.*

Camel coffee mug, traditional Old Joe design. Made in China, 1990s. 4" x 4". $10-13. *Courtesy of Odell Farley.*

Camel boxes of matchbooks. Top, 1994, bottom left, 1991, bottom right, 1995. $10-13 each. *Courtesy of Odell Farley.*

Camel Christmas ornaments. Blue and gold Camel, 1991; brass camel standing in front of tobacco leaf wreath, 1992; brass traditional camel under Advent star, 1993. $12-15 each. *Courtesy of Odell Farley.*

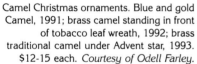

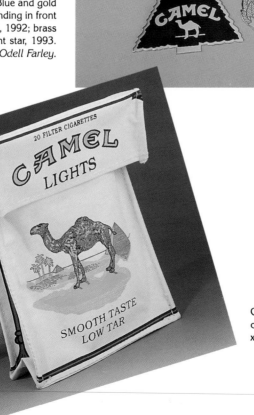

Camel fanny pack given with the purchase of a carton. The top flips up like the top of a hard pack. 6" x 5". early 1990s. $10-15. *Courtesy of Odell Farley.*

Camel sunglasses, 1990. $10-14. *Courtesy of Odell Farley.*

Brass Camel padlock keyring in ultrasuede pouch, c. 1990. $10-12. *Courtesy of Odell Farley.*

Left: Camel keyring game with the object of getting the floating cigarettes into the pack, 1990; right: small Camel cigarette pack pin, c. 1970s. $10-13 each. *Courtesy of Odell Farley.*

Pewter Camel keyring and figurine, c. 1990. $10-13 each. *Courtesy of Odell Farley.*

Three postcards for R.J. Reynolds. Two feature the rare 1915 three-panel tin display piece, and the other a camel topiary at Whitaker Park, c. 1990. $5-7 each. *Courtesy of Odell Farley.*

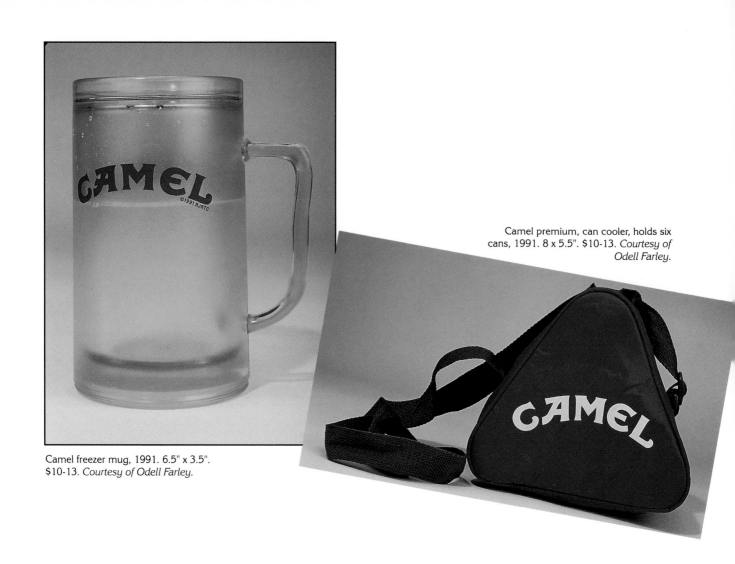

Camel freezer mug, 1991. 6.5" x 3.5".
$10-13. *Courtesy of Odell Farley.*

Camel premium, can cooler, holds six
cans, 1991. 8 x 5.5". $10-13. *Courtesy of
Odell Farley.*

"The Camel Cooler." Plastic Joe Camel
cooler mug, came with three packs,
1991. 5" x 4". $10-13. *Courtesy of Odell
Farley.*

"Camel The Hard Pack Lighters." Camel lighters given with a carton purchase. Made in France, 1991. $35-40 for the pack. *Courtesy of Odell Farley.*

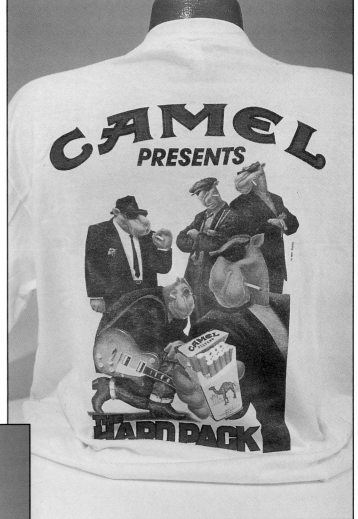

"Camel Presents the Hard Pack." T-shirt, 1991. $20-25. *Courtesy of Odell Farley.*

"Go to bat for the United Way." A shirt for RJR employees during the United Way campaign, 1991. It reads "R.J. Reynolds Tobacco Company, Whitaker Park" on back. $10-15. *Courtesy of Odell Farley.*

"Camel. Mardi Gras 91." T-shirt, 1991. On the front pocket it reads Camel Mardi Gras 1991. $10-14. *Courtesy of Odell Farley.*

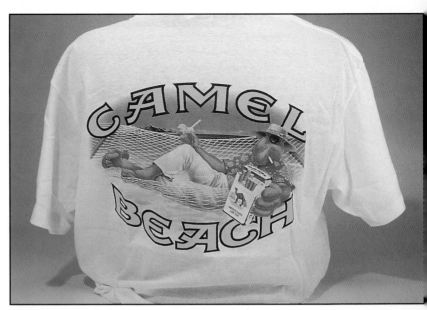

"Camel Beach." T-shirt with "Camel Daytona Beach, 1991" on the front pocket. $10-13. *Courtesy of Odell Farley.*

Camel Christmas ornament and figurine. R.J. Reynolds began putting out Christmas ornaments in 1994. Ornament, $10-14; figurine: $25-30. *Courtesy of Odell Farley.*

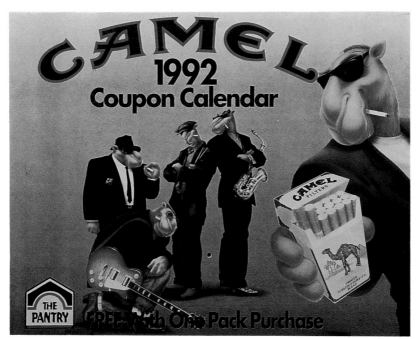

Camel 1992 Coupon Calendar, free with one pack purchase. $5-7. *Courtesy of Odell Farley.*

Camel characters shower curtain, 1991. 6' x 6'. $20-25. *Courtesy of Odell Farley.*

The Hard Pack beach towel, 1991. 57" x 28.5". $20-25. *Courtesy of Odell Farley.*

Camel Joe Pool Player beach towel, 1991, 57" x 28.5". $20-25. *Courtesy of Odell Farley.*

Advertising plates, 1991-1992. Left: Reproduction of oil painting "Camels of Course", by Frederic Mizen from 1920s to 40s. Created in limited edition for R.J. Reynolds, 1991. Marked "Fine Ivory China by World Wide." Right: Reproduction of advertisement used in 1927, created in limited edition for R.J. Reynolds, 1992. Marked "Fine Ivory China by World Wide." Numbered 32/1000. $30-35 each. *Courtesy of Odell Farley.*

Joe Camel playing cards, came with
purchase, c. 1991. $10 per the pack.
Courtesy of Odell Farley.

"Free Hard Pack Flip Top Lighter." Camel
flip top lighter, 1991. $10-13. *Courtesy of
Odell Farley.*

"Buy 3 Get 2 Free." Promotional box
featuring the Hard Pack, 1991. 4.25" x
11". $10-13. *Courtesy of Odell Farley.*

"Thank You" and "Open" and "Closed" sliding sign. Raised plastic figure of Joe leaning on a pack of cigarettes, 1991. $10-12. *Courtesy of Odell Farley.*

Push and pull store decal signs, 1991. $10-12. *Courtesy of Odell Farley.*

Camel address or business hours decal for store window, 1991. $10-12. *Courtesy of Odell Farley.*

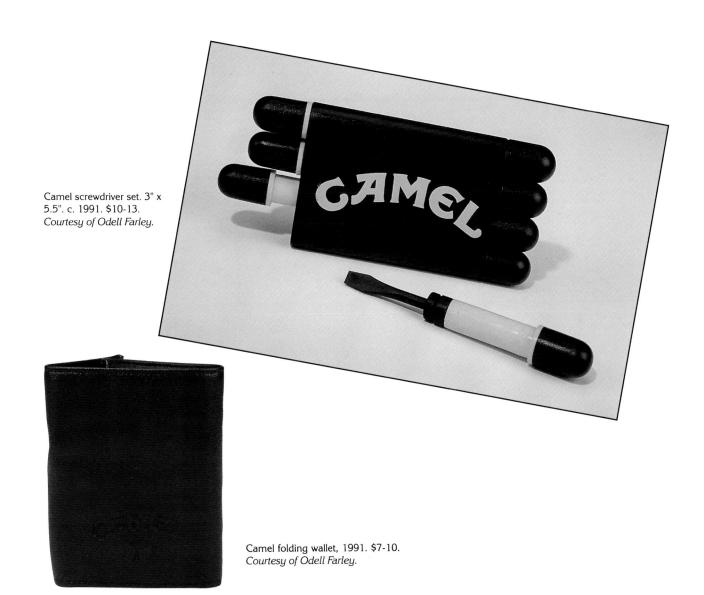

Camel screwdriver set. 3" x 5.5". c. 1991. $10-13. *Courtesy of Odell Farley.*

Camel folding wallet, 1991. $7-10. *Courtesy of Odell Farley.*

Camel net cut-off T-shirt, c. 1991. $10-15. *Courtesy of Odell Farley. $10.*

Wood and glass clock, face is the famous Camel Cigarettes triptych counter display. 13" x 9.5". c. 1991. $50-75. *Courtesy of Odell Farley.*

Camel socks, c. 1991. $10-13. *Courtesy of Odell Farley.*

Digital Camel watches, free with purchase, 1991. $10-15 each. *Courtesy of Odell Farley.*

Refrigerator magnets, c. 1991. $5-7.
Courtesy of Odell Farley.

Tins, Tales & Trademarks. A catalog of the exhibit of memorabilia, in corporate headquarters. c. 1990.

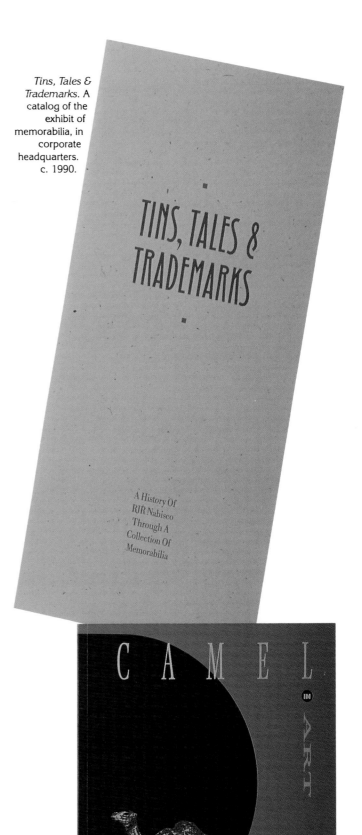

TINS, TALES & TRADEMARKS

A History Of
RJR Nabisco
Through A
Collection Of
Memorabilia

Les Aventures publicitaires d'un Dromadarie exhibit, Paris, 1992. The catalog of a Louvre Museum of Camel advertising art. 1992. $45-50.

Camel in Art. This German book was published for R.J. Reynolds, European division. Early 1990s. $30-40.

73

Camel Cash Catalog 1

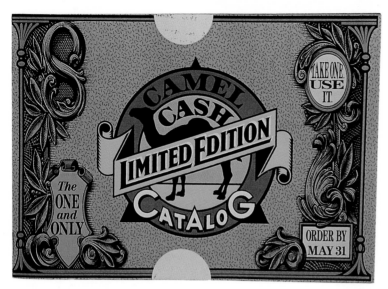

Camel Cash Catalog 1, 1991. Orders had to be received by May 31, 1992. $10-15. *Courtesy of Odell Farley.*

"Born To Be Smooth." Joe Camel T-shirt, 1991, from the first catalog. $20-25. *Courtesy of Odell Farley.*

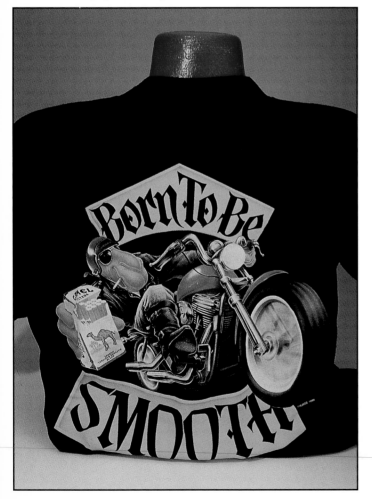

T-shirt featuring Joe Camel busting through a brick wall, 1991. The pocket is blank. It appeared in the first catalog, which expired on May 31, 1992. $20-25. *Courtesy of Odell Farley.*

Camel flip-flops with Camel Beach logo, 1991. They appeared in Camel Cash Catalog 1. $10-13. *Courtesy of Odell Farley.*

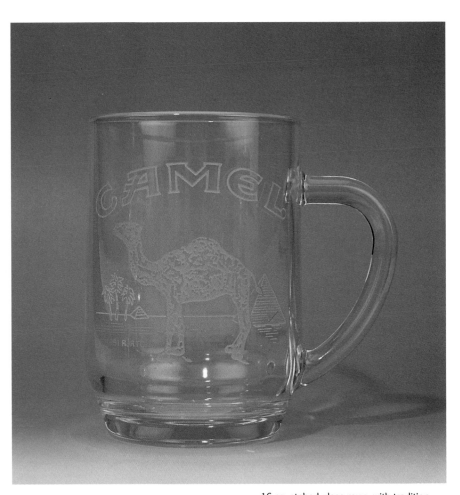

16 oz. etched glass mug, with tradition Camel logo, 1991. $10-15. *Courtesy of Odell Farley.*

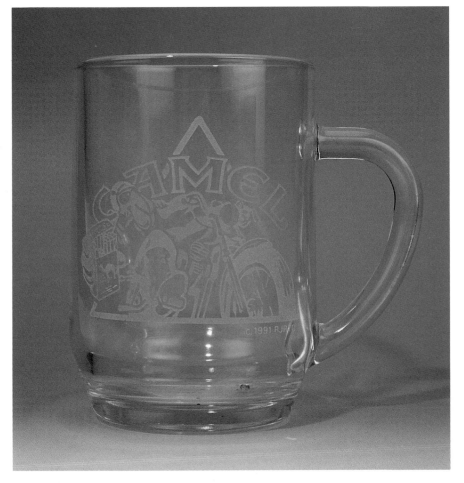

16 oz. etched glass mug, Born to be Smooth, 1991. $10-15. *Courtesy of Odell Farley.*

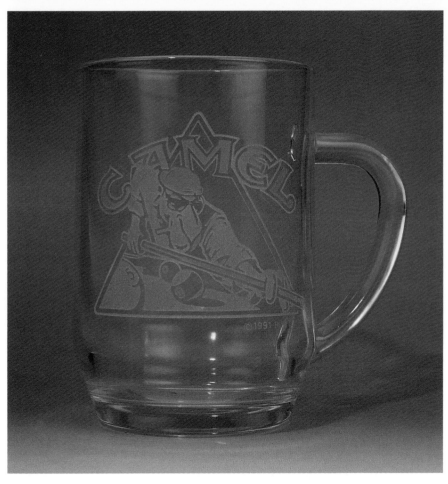

16 oz. etched glass mug, Pool Player, 1991. $10-15. *Courtesy of Odell Farley.*

Plastic Hard Pack cups featuring Joe, Floyd, Eddie, Max, and Bustah. From Catalog 1, 1991. $5-7 each. *Courtesy of Odell Farley.*

Joe Camel tie, from Catalog 1, 1991.
$10-15. *Courtesy of Odell Farley.*

Brass Zippo lighters with traditional logo
and Tuxedo Joe. Catalog 1, 1991. $35-
40 each. *Courtesy of Odell Farley.*

"Born to be Smooth" and "Hard
Pack" flip-top lighters. Catalog 1,
1991. $10-15 each. *Courtesy of
Odell Farley.*

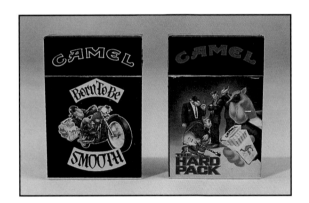

Two baseball caps: left, Pool Playing
Joe; right, Camel Beach. From
Catalog 1, 1991. $10-15 each.
Courtesy of Odell Farley.

Pool Player and Joe Camel fleece shorts. From Catalog 1, 1991. $10-15 each. *Courtesy of Odell Farley.*

Hard Pack U.S. Tour jacket. From Catalog 1, 1991. $35-50. *Courtesy of Odell Farley.*

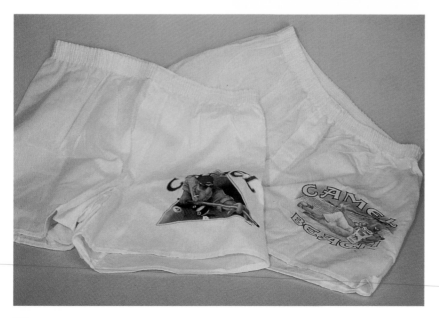

Camel Beach and Pool Player boxer shorts. From Catalog 1, 1991. $10-15 each. *Courtesy of Odell Farley.*

1992

"Club Camel." Plastic tumblers featuring Joe and the Hard Pack. Left to right: Eddie, Max, Joe, Bustah, and Floyd. 1992. 6" x 3.5". $5-7 each. *Courtesy of Odell Farley.*

Plastic Camel mugs, with lid. L to R: Playing cards, Joe playing pool, Hard Pack performing, bar scene, and boxing match, 1992. 6.5" x 4". $5-7 each. *Courtesy of Odell Farley.*

Reverse of "Club Camel" plastic tumblers.

"Club Camel." Plastic can cooler that came with two packs of cigarettes. 4" x 4". $5-7 each. *Courtesy of Odell Farley.*

Etched glass Camel pitcher and glasses, 1992. 8" x 5" and 5.75" x 2". $35-45 set. *Courtesy of Odell Farley.*

"1993 Camel Weekend Calendar." 10" x 12". $20-25. *Courtesy of Odell Farley.*

"Free with 2-Pack Purchase." A deck of Hard Pack playing cards came with two pack purchase, 1992. $15-18. *Courtesy of Odell Farley.*

"Pack-Lite III." Refillable flip-up lighter, 1992. $10-13. *Courtesy of Odell Farley.*

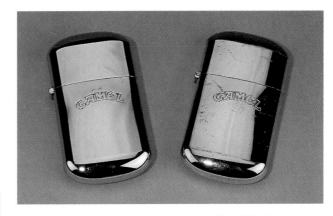

Large World War II-style trench lighter with flip-top , 1992. Colibri. $15-20. *Courtesy of Odell Farley.*

"Pack Lite III" Refillable Lighter." Camel refillable lighter featuring Joe Camel, with original box, 1992. $20-25. *Courtesy of Odell Farley.*

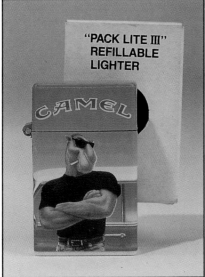

"Club Camel Lighters." Camel lighters which came with a carton of cigarettes. Made in France. 1992. $35-40 set. *Courtesy of Odell Farley.*

Camel Hard Pack coasters, 1992. Camel Lights, Joe playing pool, Camel Wides, and Hard Pack. 4" x 4". $20-25 set. *Courtesy of Odell Farley.*

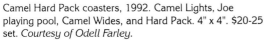

"Club Camel." Camel flip-flops, the reverse logo on the sole left a positive impression of the word Camel in the sand. $10-15. *Courtesy of Odell Farley.*

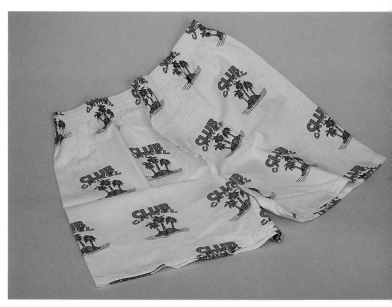

Club Camel cotton shorts. $10-15. *Courtesy of Odell Farley.*

"Club Camel, Wish You Were Here." T-shirt, free with the purchase of a three pack, 1992. $10-15. *Courtesy of Odell Farley.*

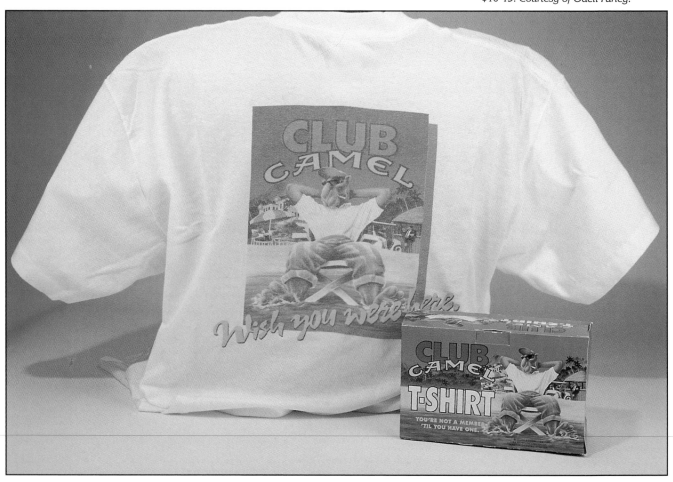

"Vote Joe." The Camel promotion during the 1992 elections ran Joe Camel for office. (Could this be the reason Bill Clinton ran so hard against cigarettes in his 1996 campaign?) The promotion produced a variety of items including a T-shirt, buttons, ashtrays, bumper stickers, and paper cups. T-shirt: $10-15; $3-5 each for everything else. *Courtesy of Odell Farley.*

"Boston." T-shirt featuring Joe in a Boston sweat shirt with hockey gear, 1992. $10-15. *Courtesy of Odell Farley.*

"Vote and may the best party win!" Paper cup from the election year promotion, 1992. 3" x 4.5". $5-7. *Courtesy of Odell Farley.*

"Go to Bat for the United Way." T-shirt manufactured for RJR employees to promote donations to the United Way. Joe Camel has an RJR baseball uniform, with the United Way logo on the bat. The back of the shirt reads "R.J. Reynolds Tobacco Company, Whitaker Park." $10-15. *Courtesy of Odell Farley.*

"Felices 500, Americas!" Spanish language T-shirt commemorating Columbus's discovery of the Americas. The same graphics are on the front pocket. $10-15. *Courtesy of Odell Farley.*

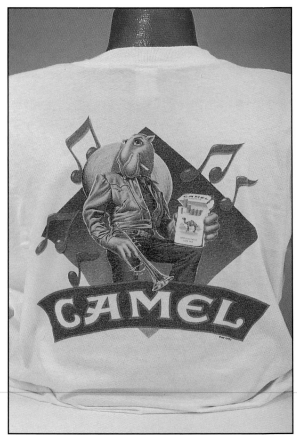

"Viva La Independencia!" Joe Camel with a sombrero and trumpet grace this T-shirt, 1992, celebrating Mexican Independence day. On the front pocket it reads "!Que Viva La Independencia!". $10-15. *Courtesy of Odell Farley.*

"Camel." T-shirt featuring Joe Camel on motorcycle, 1992. On the front pocket it reads "Daytona Beach Bike Week '92." $10-15. *Courtesy of Odell Farley.*

Wides Guys fleece shorts, free with three pack purchase, 1992. $15-20. *Courtesy of Odell Farley.*

Camel Wides windbreaker, free with three packs, 1992. $20-25. *Courtesy of Odell Farley.*

Camel Wides playing cards, came with purchase, 1992. $10-13. *Courtesy of Odell Farley.*

Joe's Tackle Shop tackle box. $20-25.
Courtesy of Odell Farley.

Club Camel Official
Playing Cards, came with
purchase, 1992. $10-13.
Courtesy of Odell Farley.

Camelflage Joe T-shirt, 1992. $10-15.
Courtesy of Odell Farley.

Camelflage belt, 1992. Note Camel
pattern on belt fabric. $10-15.
Courtesy of Odell Farley.

Thank You, Call Again sign for store window, 1992. $10-13. *Courtesy of Odell Farley.*

"Camel The Year in Pictures," 1993 calendar. $10-13. *Courtesy of Odell Farley.*

Joe Camel clock, 1992. Plastic. $25-30.
Courtesy of Odell Farley.

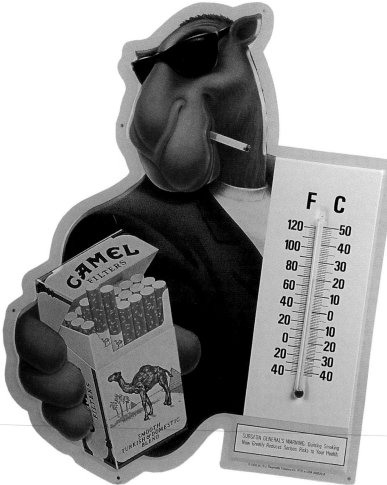

Joe Camel thermometer, 1992. Plastic.
$25-30. *Courtesy of Odell Farley.*

"Joe Camel and The Hard Packs Latest
Gig...Arts Council, Winston-Salem."
Community service poster, c. 1992. $20-
25. *Courtesy of Odell Farley.*

"New Camel Wides." Translite featuring
Max and Ray, the Wides Guys, 1992. $25-
30. *Courtesy of Odell Farley.*

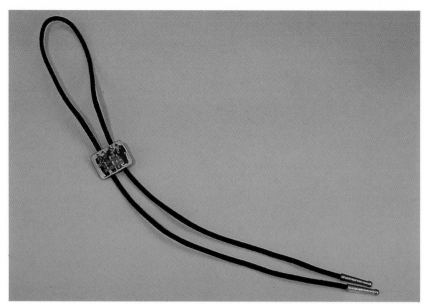

Bolo tie with Hard Pack slide, c. 1992. $10-14. *Courtesy of Odell Farley.*

Pool Player auto shade. 21" x 52.5". c. 1992. $10-13. *Courtesy of Odell Farley.*

Camel Cash: The Catalog, Volume II

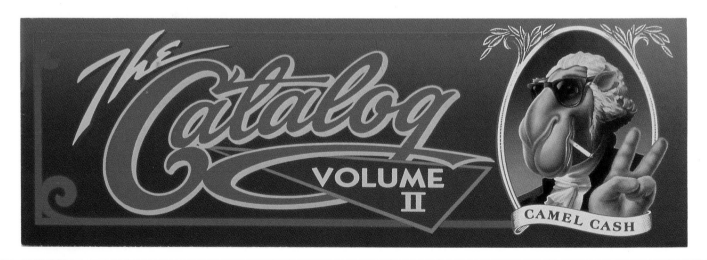

Catalog II, 1992. Orders had to be received by January 1, 1993. $10-15. *Courtesy of Odell Farley.*

Gold-plated sterling silver Camel stud earrings. 0.5" x 0.5". From Catalog 2, 1992. $25-35. *Courtesy of Odell Farley.*

Pack of five Club Camel lighters featuring Floyd, Eddie, Joe, Max, and Bustah. From Catalog 2, 1992. $35-45 for the pack. *Courtesy of Odell Farley.*

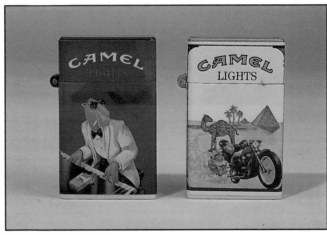

Two refillable butane lighters: Piano Player and Desert Biker. Catalog 2, 1992. $20 each. *Courtesy of Odell Farley.*

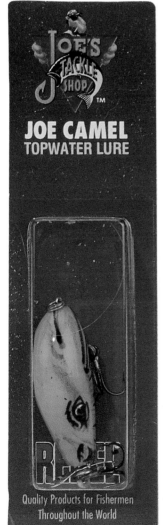

Cobalt blue half-moon glass ashtray, 7.8" x 3.75". From Catalog 2. $20-25. *Courtesy of Odell Farley.*

"Joe's Tackle Shop," Genuine Rebel Joe Camel Top Water Lure, 2.25". Catalog 2, 1992. $10-15. *Courtesy of Odell Farley.*

Five postcards. Left to right: Joe in white dinner jacket, the Hard Pack, Joe on motorcycle, Joe shooting pool, and Joe at Club Camel. Packed in a set of ten cards from Catalog 2, 1992. $3-4 each. *Courtesy of Odell Farley.*

Left: Camel Java Mug, 11 oz., ceramic. Right: Joe's Mug, heavy plastic, 14 oz. From Catalog 2, 1992. $5-7 each. *Courtesy of Odell Farley.*

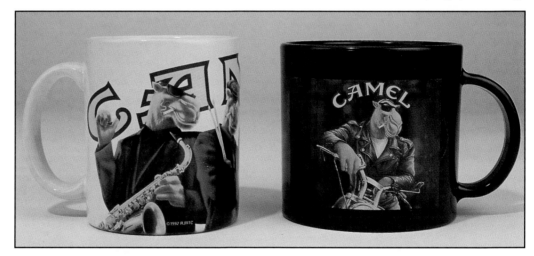

Three Camel Collector's Tins with fifty books of matches inside of each. Classic Camel, Deco Camel, and Pool Player. 7" x 4.5". From Catalog 2, 1992. $15-20 each. *Courtesy of Odell Farley.*

Swiss army knife, 2.25", and two midnight chrome Zippo lighters, Biker and classic Camel. Catalog 2, 1992. Zippo lighters, $35-40; knife, $15-20 each. *Courtesy of Odell Farley.*

Plastic Camel steins, 24 oz. Hard Pack, Club Camel, Wides, and Golf. From Catalog 2, 1992. $5-7 each. *Courtesy of Odell Farley.*

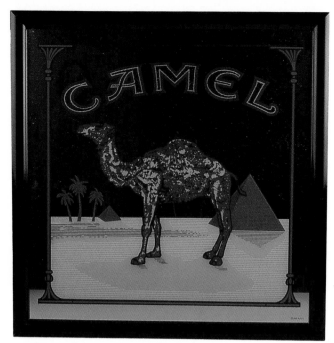

"Classic Camel." Logo mirror, 1992. Part of a series of five offered in Catalog 2. Series of five. $35-40 each. *Courtesy of Odell Farley.*

"Piano Player." Mirror. From Catalog 2, 1992. *Courtesy of Odell Farley.*

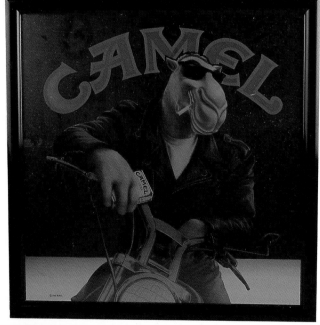

"Biker." Joe on motorcycle mirror. From Catalog 2, 1992. *Courtesy of Odell Farley.*

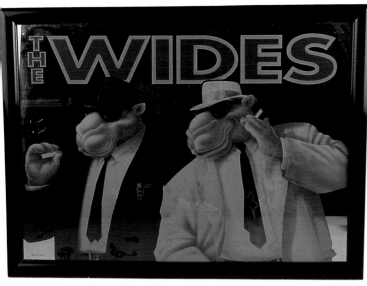

"The Wides." Mirror. From Catalog 2, 1992. *Courtesy of Odell Farley.*

"The Hard Pack." Mirror. From Catalog 2, 1992. *Courtesy of Odell Farley.*

Tropical Shower Curtain, 6' x 6'. From Catalog 2, 1992. $20-25. *Courtesy of Odell Farley.*

License plates in English and Spanish: Smooth 1 and Suave 1, 1992, found in Catalog 2. $10-12 each. *Courtesy of Odell Farley.*

1993

Plastic Camel mugs. Left: Camel GT racing, Right: Atlanta Knights, professional hockey. 1993. $5-7 each. *Courtesy of Odell Farley.*

"Smooth Character." Joe Camel travel mugs, insulated, made by Aladdin. The two on the left are the same. All after 1993. 7" x 5". $5-7 each. *Courtesy of Odell Farley.*

St. Patrick's Day plastic mug featuring
Max and Ray, "The Wide Guys," 1993.
6.5" x 4". $10-13. *Courtesy of Odell
Farley.*

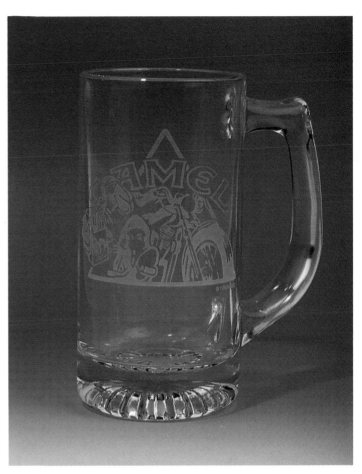

Glass Camel mug, Joe on a motorcycle.
Copyright 1993. 3.5" x 6". $10-12.
Courtesy of Odell Farley.

Old fashioned glasses,
"Smooth Character," 1993.
3.5" x 3.5". $5-7 each.
Courtesy of Odell Farley.

Etched Camel shot glasses, 1993. Left to right: Joe on a motorcycle, traditional Old Joe Camel design, and Joe playing pool. $5-7 each. *Courtesy of Odell Farley.*

"Joe's Fish & Game Club." Enameled camping mug, 1993. $10-13. *Courtesy of Odell Farley.*

Camel Cash lighters, Made in France, 1993. $8-10 each. *Courtesy of Odell Farley.*

Camel tin, 1993. Made in Hong Kong. 4" x 9.5". $20-25. *Courtesy of Odell Farley.*

Close-up of Camel tin graphics.

"Holiday Lighters." Five lighters with the Hard Pack in holiday dress came with a carton purchase. Made in France, 1993. $35-40 for the pack. *Courtesy of Odell Farley.*

"Play Camel Cash Lotto." Front and back of Camel Cash Lotto lighter, 1993-1994. $8-10 each. *Courtesy of Odell Farley.*

"Joe's on the Beach" coasters, 1993. Eddie, Max, Floyd, and Bustah. 4" x 4". $15-18. *Courtesy of Odell Farley.*

Camel Hard Pack CD carrier, 1993. A gift with the purchase of a carton. 6" x 5.5". $5-7. *Courtesy of Odell Farley.*

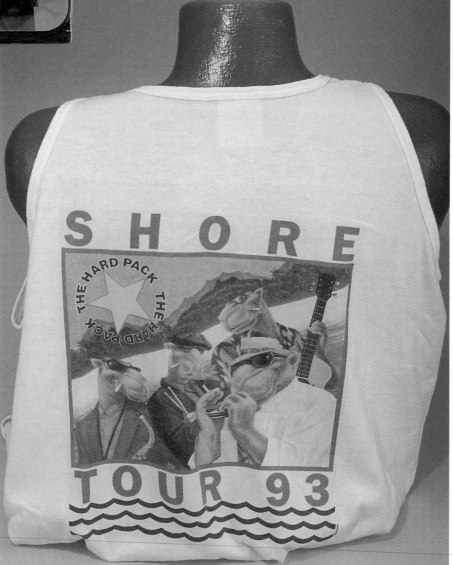

The Hard Pack "Shore Tour 93" tank top, on front is "Camel - Rhode Island." $10-13. *Courtesy of Odell Farley.*

Camel tank top, Joe Camel on back, "Camel Chicago 1993" on front. $10-13. *Courtesy of Odell Farley.*

Matching Joe Camel keychains, pins, and money clip, c. 1993. $3-5 each. *Courtesy of Odell Farley.*

Flip-top box of Camel filter wooden matches, and a slide-open box of Motorsports matches, 1993. $3-5 each. *Courtesy of Odell Farley.*

Gold-colored Joe Camel dog tags, limited edition numbered 0004. 1993. $10-13. *Courtesy of Odell Farley.*

Vinyl counter pad with rubber backing, featuring Joe, 1993. $10-13. *Courtesy of Odell Farley.*

Camel Special Light notepad, advertisement, and pack of matches, 1993. $10-13. *Courtesy of Odell Farley.*

Top left, 1995 design gold plated "Genuine Taste" pin; top right: 1993 Thank you pin and card to Camel smokers; bottom left, silver and gold-colored zipper pulls; bottom right a brass keyring. $4-6 each. *Courtesy of Odell Farley.*

Camel Cash Catalog 3

"Camel." Joe's Garage Tee, 1993, from Catalog 3. The front pocket has Joe's Garage logo. $10-15. *Courtesy of Odell Farley.*

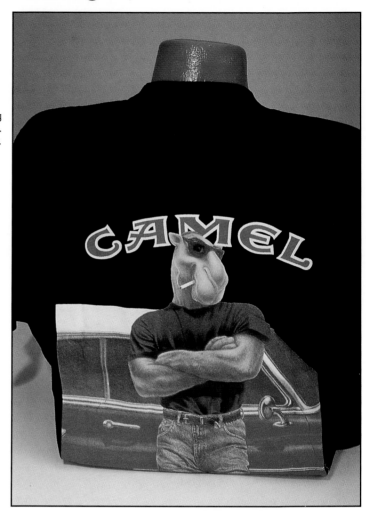

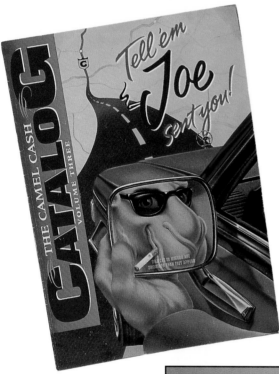

Catalog 3, late 1992 to 1993. Orders needed to be received by July 31, 1993. $10-13. *Courtesy of Odell Farley.*

A variety of items from Catalog 3. Back left: Road Cup and Holder. Front, left to right: Joe's Garage key ring; The Wides and Joe Camel flip-top refillable lighters; Pool Player cigarette tin; five pack of Smooth Deal lighter set. 1993. Lighters, $5-12; 5-pack, $35-40; all others, $5-7 each. *Courtesy of Odell Farley.*

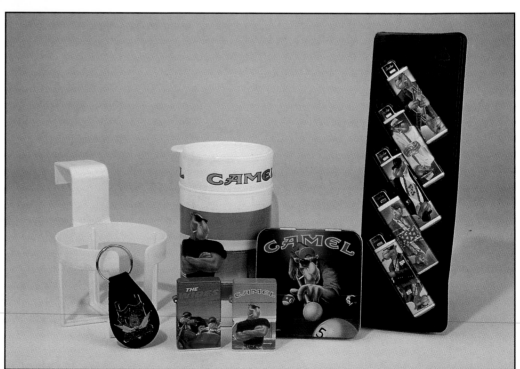

Neoprene "Hammock Hugger," went over 12 oz. bottles for insulation. From Catalog 3, 1993. $8-12. *Courtesy of Odell Farley.*

"Tropical Hugger," featuring the Hard Pack. Catalog 3, 1993. $8-12. *Courtesy of Odell Farley.*

Max and Ray plastic salt and pepper shakers, 4" high. From Catalog 3, 1993. $10-15. *Courtesy of Odell Farley.*

Left: Camel Cash Clip, 2.5" x 1.5". Right: Joe's Diner Deco Chrome Lighter, refillable. From Catalog 3, 1993. $10-13 each. *Courtesy of Odell Farley.*

The Camel Game, with forty-eight cards and six dice. From Catalog 3, 1993. $10-13. *Courtesy of Odell Farley.*

Classic Camel ashtray, 5"
each side; Classic Camel
cigarette tin; and Camel
"Lights" flashlight/key
chain. From Catalog 3,
1993. $5-7 each. *Courtesy
of Odell Farley.*

Floating key chain/refillable
lighter and Outdoor Zippo.
From Catalog 3, 1993.
$30-35 each. *Courtesy of
Odell Farley.*

Joe's Diner Jukebox. Tabletop
jukebox replica, AM/FM radio and
cassette player. 13" x 12" x 5". From
Catalog 3. $50-65. *Courtesy of Odell
Farley.*

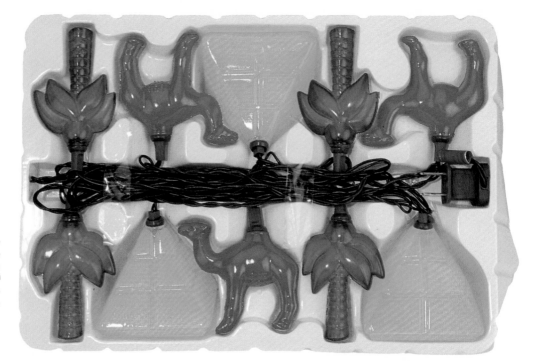

Party lights, ten bulbs, including Camels, pyramids, and palm trees, 14 feet long. From Catalog 3, 1993. $20-25. *Courtesy of Odell Farley.*

Joe's On the Beach mesh tank-top. From Catalog 3, 1993. $10-13. *Courtesy of Odell Farley.*

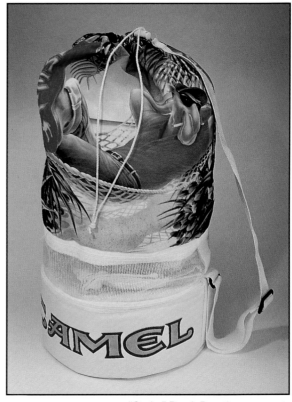

"Cooler" Beach Bag, three compartments, 25" x 12". From Catalog 3, 1993. $10-13. *Courtesy of Odell Farley.*

Hard Pack BBQ Apron with Joe's Diner on pocket. From Catalog 3, 1993. $10-13. *Courtesy of Odell Farley.*

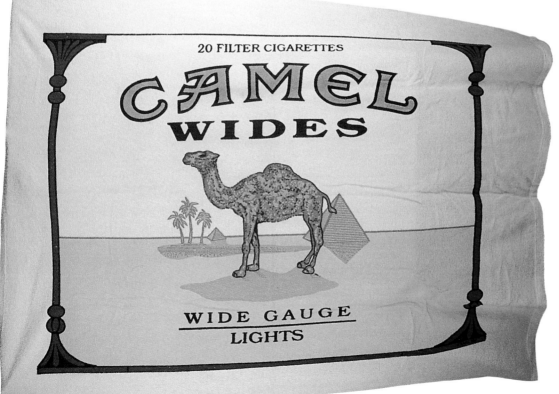

Camel Wides double size beach towel. From Catalog 3, 1993. $12-15. Courtesy of Odell Farley.

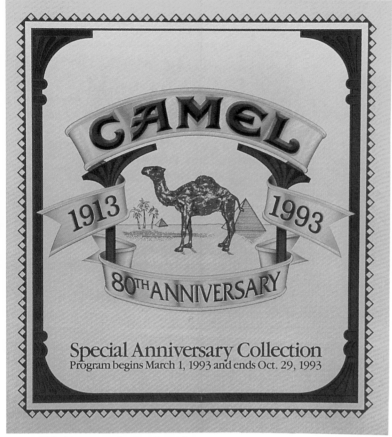

A close-up of the poker chip showing the classic Camel logo on one side and the 80th anniversary logo on the other.

80th Anniversary Catalog, Special Anniversary Collection, March 1, 1993 to October 29, 1993. Items in this catalog were bought with UPC proofs of purchases and money. The catalog was aimed at the smokers of the original Camel non-filter cigarettes. $10-15. *Courtesy of Odell Farley.*

Limited Edition Poker Set contains two packs of Camel playing cards and 150 poker chips, in a tin decorated with vintage advertising. Made in England. From the 80th Anniversary Catalog. $30-40. *Courtesy of Odell Farley.*

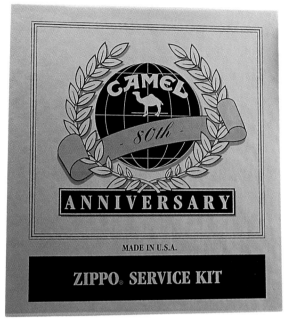

Though not in the Anniversary Catalog, this Camel Zippo Service Kit was made for the 80th Celebration. It includes a chrome Zippo with a pewter logo, extra flints, lighter fluid, tweezers, and a brush for cleaning, all in commemorative box. The 80th Anniversary logo on the lighter is different from that used in the catalog, so this may be a prototype for a logo design that was rejected. $40-45. *Courtesy of Odell Farley.*

Brass Camel Heritage Ashtray, 5" cast metal with antique finish. From the 80th Anniversary Catalog. $10-13. *Courtesy of Odell Farley.*

A card box in the size and form of a Camels Fifties box with two decks of poker-sized playing cards. From the 80th Anniversary Catalog. $20-25. *Courtesy of Odell Farley.*

Left: Camel-pack lighter; right: brass lighter with Camel 1913-1993 logo. From the 80th Anniversary Catalog. $10-15 each. *Courtesy of Odell Farley.*

111

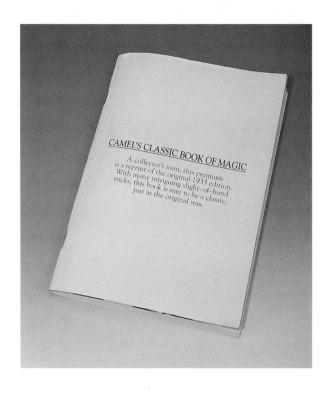

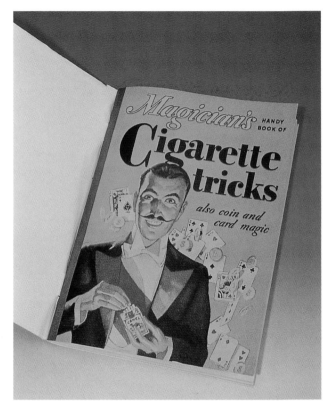

Reprint of the 1933 *Magician's Handy Book of Magic Tricks* with extra new cover. From the 80th Anniversary Catalog. $5-7. *Courtesy of Odell Farley.*

Pocket knives with steel blades. Left, wood-handled engraved pocket knife; right, solid brass pocket knife. From the 80th Anniversary Catalog. $20-25 each. *Courtesy of Odell Farley.*

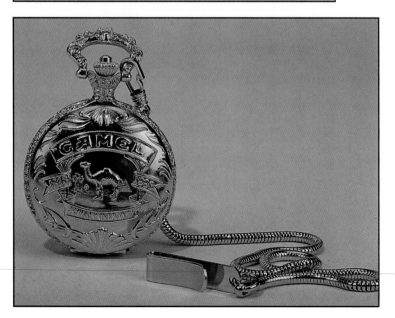

Anniversary Pocket Watch, 24k gold plating with chain. Shown closed and opened. From the 80th Anniversary Catalog. $35-45. *Courtesy of Odell Farley.*

Anniversary shot glass. This came in a set of two. From the 80th Anniversary Catalog. $3-5 each. *Courtesy of Odell Farley.*

80th Anniversary Catalog mirror, 1993. $20-25. *Courtesy of Odell Farley.*

Motorsports Catalog, 1993

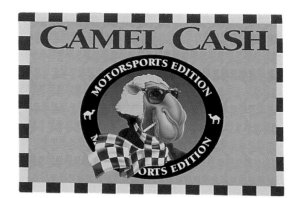

Flip-top Team Camel lighters, Supercross, Pro, and GT, with mini-Motorsport catalog. 1993. $10-14 each. *Courtesy of Odell Farley.*

Set of eight Camel blazer buttons, two large and six sleeve sizes. From Motorsports catalog. $10-15 set. *Courtesy of Odell Farley.*

Six Camel Motorsport pins. $5-7 each. *Courtesy of Odell Farley.*

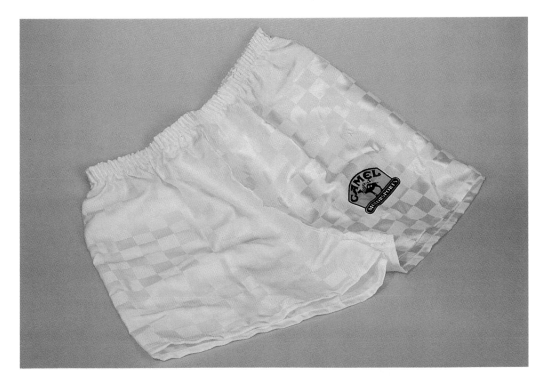

Camel Motorsports
nylon shorts, 1993.
$5-8. *Courtesy of
Odell Farley.*

Camel sunglasses from
Motorsports Collection,
1993. $7-12. *Courtesy
of Odell Farley.*

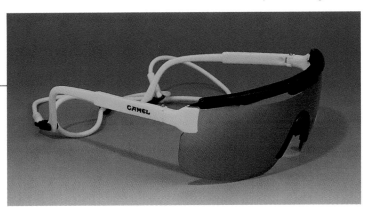

Camel Cash Catalog 4, 1993

Catalog 4 and related materials which
introduced the Camel Cash Lotto. 1993.
Orders needed to be received by February
29, 1994. $10-13. *Courtesy of Odell
Farley.*

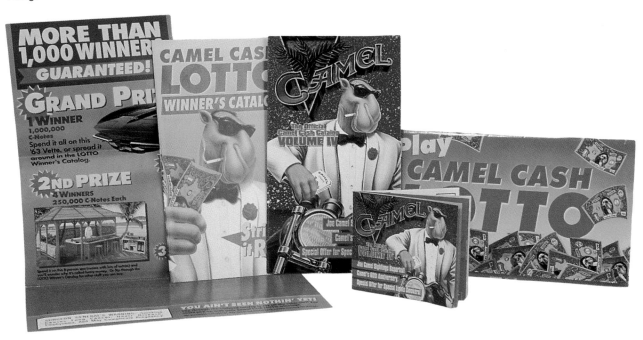

Diamond cobalt-blue Camel Special Lights ashtray, 4.5" x 8". From Catalog 4, 1993. $5-7. *Courtesy of Odell Farley.*

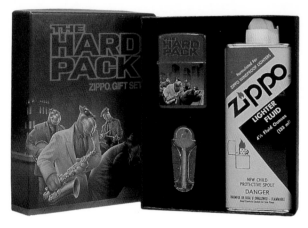

"Meet the Hard Pack" cassette, and Hard Pack Zippo Gift Set which came with a Hard Pack Zippo, spare flints, and lighter fluid. From Catalog 4, 1993. $40-45 set. *Courtesy of Odell Farley.*

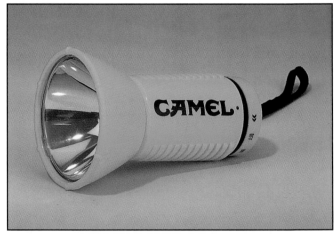

Two-in-one flashlight/lantern, 5.25". Catalog 4, 1993. $10-12. *Courtesy of Odell Farley.*

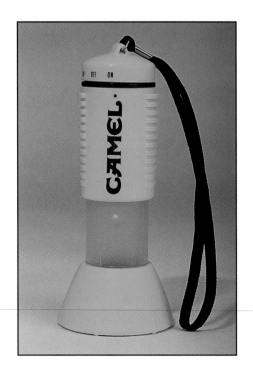

116

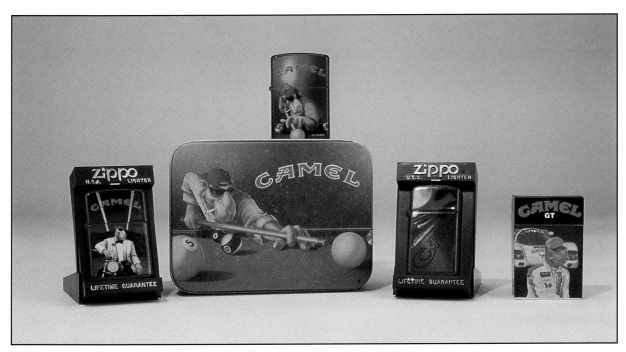

From Catalog 4, 1993, left to right: Joe
Camel White Tux Zippo; Pool Player Zippo
and tin set; the Special Slimline Zippo;
and butane racing car butane lighter (not
in Catalog 4). $5-7 butane, $30-35
Zippos. *Courtesy of Odell Farley.*

Max and Ray Comin' and Goin' Ashtray, glass ashtray
with two sides, 4.75" x 5.75". From Catalog 4, 1993.
$5-8. *Courtesy of Odell Farley.*

Camel C-handle plastic mug, holds 14 oz.
Catalog 4, 1993. $5-8. *Courtesy of Odell
Farley.*

Gold-plated, sterling silver necklace and
earrings. The necklace has a 20" chain.
Catalog 4, 1993. $20-25. *Courtesy of
Odell Farley.*

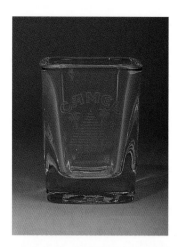

Camel Shot Glass, 2.5 oz, square with etched Camel logo. Catalog 4, 1993. $5-7. *Courtesy of Odell Farley.*

Pyramid shaped quartz analog alarm clock, with "sleepy camel" graphics and glow-in-the-dark hands. Plastic, 4.5" x 4.5". Catalog 4, 1993. $20-25. *Courtesy of Odell Farley.*

Front: heavy metal nickel-plated lighter case; back: Camel Windproof Lighter, with special heating coil. From Catalog 4, 1993. $10-13 each. *Courtesy of Odell Farley.*

Special Collector's Match Tin, limited edition, it came with fifty books of matches. 7" x 4.5". Catalog 4, 1993. $10-12. *Courtesy of Odell Farley.*

Joe Camel Key Chain, pewter medallion on a leather strap. From Catalog 4, 1993. $5-8. *Courtesy of Odell Farley.*

Classic Brass Box with hand-cutout graphics and removable lid, 4" x 7". Catalog 4, 1993. $15-20. *Courtesy of Odell Farley.*

Silver 1 oz. Camel Commemorative Coin for the 80th birthday of Camels. It came in a suede pouch. Catalog 4, 1993. $20-25. *Courtesy of Odell Farley.*

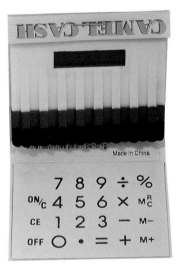

Camel Cash matchbook calculator. From Catalog 4, 1993. $10-13. *Courtesy of Odell Farley.*

119

Pool Player Wall Clock, with molded face
and three-dimensional balls. 10.5"
diameter, Catalog 4, 1993. $20-25.
Courtesy of Odell Farley.

Camel "Light," "permanent match."
From Catalog 4, 1993. $5-8. *Courtesy
of Odell Farley.*

Colossal Camel Matches, a 15" long
matchbook, containing 240 matches.
Catalog 4, 1993. $5-7. *Courtesy of Odell
Farley.*

1994

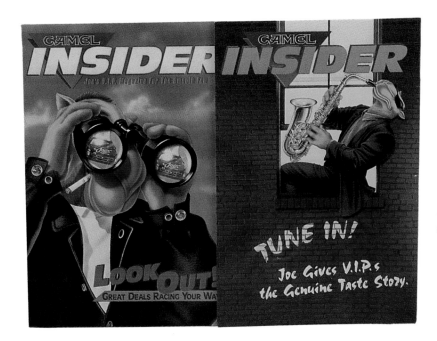

"Camel Insider," first editions. 1994. $5-8.
Courtesy of Odell Farley.

Camel Classifieds 1995, 1995 VIP Camel
calendar. $5-8. *Courtesy of Odell Farley.*

Maglite keychain flashlight, 1994. 3.5".
$7-10. *Courtesy of Odell Farley.*

Plastic Camel tumblers. Left to right: Hard
Pack, Joe's Place, and Dallas Alley Club
Camel, c. 1994. 4.5" x 3.5". $5-7 each.
Courtesy of Odell Farley.

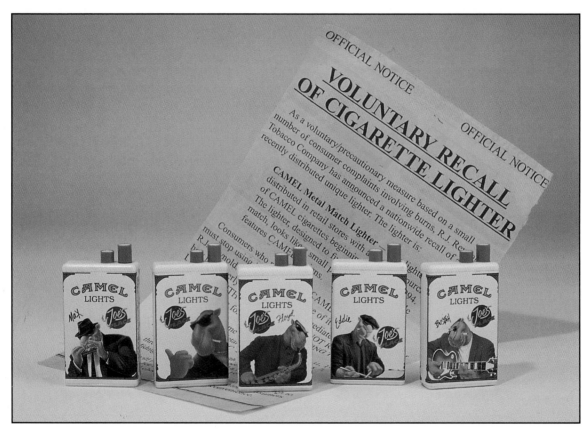

"Camel Lights." Camel metal match
lighters, featuring the Hard Pack and
Joe's Place, 1994. They were voluntarily
recalled because the fluid leaked. Left to
right: Max, Joe, Floyd, Eddie, and Bustah.
1.5" x 2.5". $20-25 each. *Courtesy of
Odell Farley.*

Camel Colibri Electro-Quartz lighter,
1994. $10-14. *Courtesy of Odell Farley.*

Black Camel Genuine Taste tin box, 1994.
4.5" x 2". $5-8. *Courtesy of Odell Farley.*

Catalog 5, 1994

"Joe's Place," Catalog 5, 1994. The last
date for orders was December 31, 1994.
$5-10. *Courtesy of Odell Farley.*

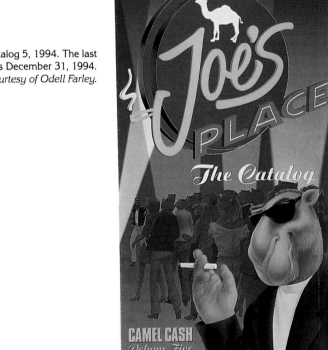

Tropical Hard Pack Tin & Lighter
Ensemble, 4.5" x 12". Catalog 5, 1994.
$35-40. *Courtesy of Odell Farley.*

Hard Pack poker set, with two Camel card decks, poker chips, and removable chip and card caddie, in a tin box. 5.5" x 5.5". From Catalog 5, 1994. $30-40. *Courtesy of Odell Farley.*

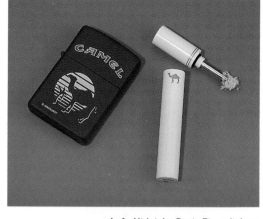

Left: Midnight Oasis Zippo lighter, refillable; right, Special Lights Permanent Match, 3.25". Catalog 5, 1994. Zippo: $30-35; match: $4-8. *Courtesy of Odell Farley.*

Maple cigarette holder, 4.5" x 2.75". Catalog 5, 1994. $5-10. *Courtesy of Odell Farley.*

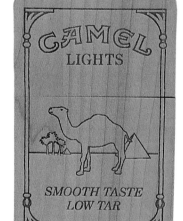

Midnight Oasis Walnut Case, with hinged top and felt lining. 3.5" x 5.5". Catalog 5, 1994. $5-10. *Courtesy of Odell Farley.*

Left to right: Joe's Place playing cards and tin, square Joe's Place shot glass, and Midnight Oasis tin. Catalog 5, 1994. $4-8 ea. *Courtesy of Odell Farley.*

Joe's Place dart board set. 18" regulation boar bristle dart board in hickory cabinet. 21.75" x 19.5" x 3". $50-75. Darts came separately. $10-15. *Courtesy of Odell Farley.*

8-Ball Smoking set which includes a windproof Zippo with an 8-ball emblem, removable floating ashtray, and 8-ball tin. 4.25". Catalog 5, 1994. $35-40. *Courtesy of Odell Farley.*

Left: Classic Trench Lighter. A brass replica of a World War One lighter, 3" x 1"; right: 1940s Style Lighter, 1.5" x 2". Catalog 5, 1994. $10-15 ea. *Courtesy of Odell Farley.*

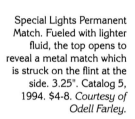

Special Lights Permanent Match. Fueled with lighter fluid, the top opens to reveal a metal match which is struck on the flint at the side. 3.25". Catalog 5, 1994. $4-8. *Courtesy of Odell Farley.*

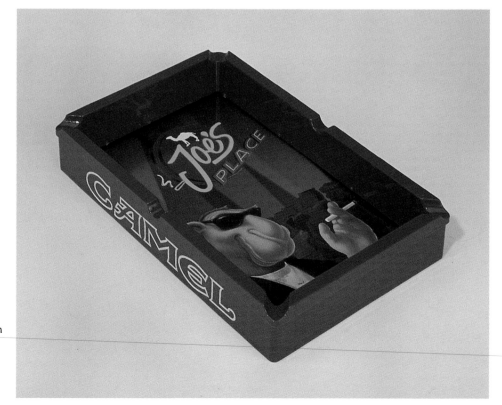

Joe's Place ashtray, plastic with glass liner. 7.75" x 4.5". From Catalog 5, 1994. $5-10. *Courtesy of Odell Farley.*

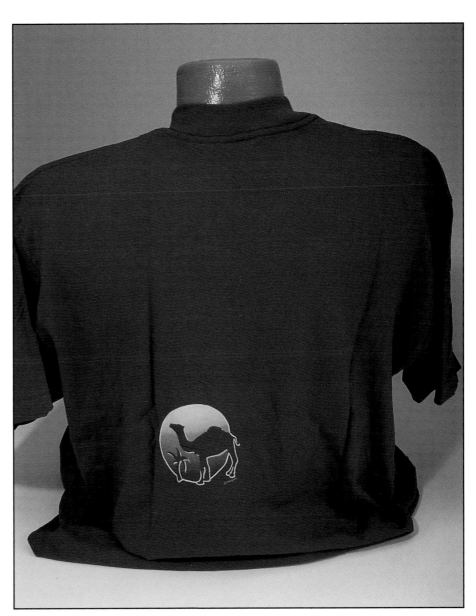

"Midnight Oasis T-shirt." A unique Camel logo appears on the back, and the pocket has a blue-green "Camel." 1994. Offered in catalog five. $10-15. *Courtesy of Odell Farley.*

Chrome plated items, left to right: Special Lights Deco lighter and cigarette case; and Camel Compact. Catalog 5, 1994. $10-15 each. *Courtesy of Odell Farley.*

Smokin' Joe's Racing, 1994

Smokin' Joe's Racing, Camel Powered 1993-94 Catalog. Orders were due by December 31, 1994. $4-6. *Courtesy of Odell Farley.*

Camel Powered Smokin' Joe's Racing shot glass and stirrer. The shot glass is in the 1994 Smokin' Joe's Racing Catalog. $4-10. *Courtesy of Odell Farley.*

Smokin' Joe's Racing Zippo and tin set, 1994, and tin match set, 1995. Not in catalog. Zippo, $35-40; Tin match set, $10-15. *Courtesy of Odell Farley.*

"Top Fuel Car" design T-shirt, 1994, Sports Marketing Enterprises. Smokin' Joe's Racing logo on front pocket. Appears in 1994 Smokin' Joe Racing catalog. $10-15. *Courtesy of Odell Farley.*

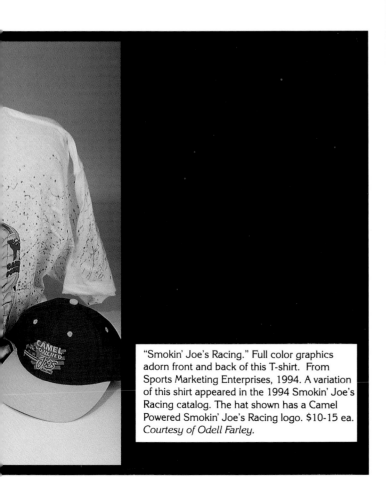

"Smokin' Joe's Racing." Full color graphics adorn front and back of this T-shirt. From Sports Marketing Enterprises, 1994. A variation of this shirt appeared in the 1994 Smokin' Joe's Racing catalog. The hat shown has a Camel Powered Smokin' Joe's Racing logo. $10-15 ea. *Courtesy of Odell Farley.*

Postcard announcing race, racing schedule, and embroidered Smokin' Joe's racing patch, 1994. Not in catalog. $2-4. *Courtesy of Odell Farley.*

Smokin' Joe's Racing Pin set, 1994, pewter, with six pins: Hut Stricklin, Jim Head, "Gordie" Brown, Mike Hale, Kevin Magee, and Mike Smith. Pins are 3" x 1.25" each. Not in catalog. $10-20 set. *Courtesy of Odell Farley.*

Smokin' Joe's Racing scarf, purple camel silhouette wearing checkered racing scarf, 1994. $10-15. *Courtesy of Odell Farley.*

Photo of Joe's Racing Nascar #23. $5-10. *Courtesy of Odell Farley.*

Left: set of five lighters which came with a carton, GT, Pro, Lights, Supercross, and Mud & Monster, 1993. Right: Smokin' Joe's Racing Lighter set, five lighters, 1994. $20-25 set. *Courtesy of Odell Farley.*

131

Tin with matches, free with three-pack
purchase, postcard, and VIP card, 1994.
$4-8. *Courtesy of Odell Farley.*

"Smokin' Joe's Racing." Butane lighter,
featuring car number 23, 1994. $4-6.
Courtesy of Odell Farley.

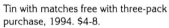

Tin with matches free with three-pack
purchase, 1994. $4-8.

"Camel Powered." T-shirt, 1994, with "Smokin' Joe's Racing" on the front pocket. $10-15. *Courtesy of Odell Farley.*

"Camel Powered" T-shirt, 1994. Smokin' Joe's Racing on front pocket. $10-15. *Courtesy of Odell Farley.*

Smokin' Joe's Racing polo shirt, 1994. Marked "Sales" on sleeve, it was worn by staff members. Beside it is a baseball cap with the same camel and scarf logo, and "Camel Powered" on back. Shirt: $15-20; cap: $10-15. *Courtesy of Odell Farley.*

133

1995

Pack display. $4-6.

"Buy 3 Packs...Get 3 free." Promotional sign, 1995. $4-6.

"Camel Free Tee Shirt." Promotional sign for a free offer with five packs, 1995. $4-6.

134

"Camel Free Lighter." Promotional sign for a free lighter with two packs, 1995. $4-6.

"Camel Free Lighter." Promotional sign for a free lighter with two packs, 1995. $3-5.

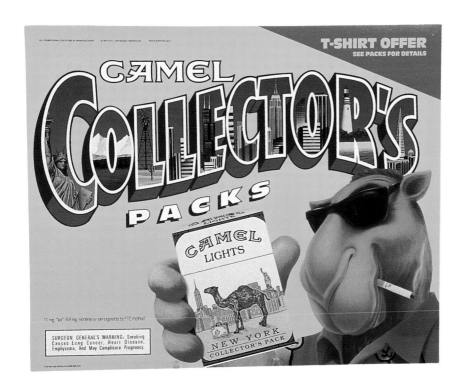

"Camel Collector's Packs." Promotional sign for specially designed state packs, 1995. $5-7.

The Camel Company Catalogs for 1995. Items in these catalogs are cash only. $2-4 ea. *Courtesy of Odell Farley.*

Diner-style ashtray from the 1995 Camel Company catalog. 1.8" x 4". $8-10. *Courtesy of Odell Farley.*

Ashtray, black with white logo, "Camel Genuine Taste," 1995. 1.8" x 4". 1995. $4-6. *Courtesy of Odell Farley.*

"Camel Powered." T-shirt with scarved racing camel logo. Copyright 1995, Sports Marketing Enterprises. "Smokin' Joe's Racing" is on the front pocket. $10-15. *Courtesy of Odell Farley.*

Camel Genuine Taste tank-top, free with three packs. $8-10. *Courtesy of Odell Farley.*

A set of five Camel Art pins from the early 1990s. Enamel on metal, marked Deco Pub, Bern. $30-40 set.

137

Camel Cash Catalog 6, 1995

Camel Cash Catalog Six, 1995. Orders due by January 31, 1996. $2-3. *Courtesy of Odell Farley.*

"Joe's Black Biker T-Shirt." Genuine Camel, 1995. The pocket has a purple and orange camel silhouette and reads "Genuine Taste." Appeared in Catalog 6. $10-15. *Courtesy of Odell Farley.*

Joe's Oktoberfest Collector's Stein, limited edition and signed, pewter and ceramic stoneware. 7.5" high, 18 oz. Catalog 6, 1995. $25-35. *Courtesy of Odell Farley.*

Classic Camel Zippo Tin set, with lighter and tin. 2" x 4.75". Catalog 6, 1995. $35-40. *Courtesy of Odell Farley.*

Hard Pack Match Cube. Acrylic cube with twenty-eight boxes of wooden matches. From Catalog 6, 1995. $4-8. *Courtesy of Odell Farley.*

Genuine Taste Shot glass. Pewter and glass, 2.5 oz. Catalog 6, 1995. $3-4. *Courtesy of Odell Farley.*

Left to right: Black Crackle Zippo lighter; Military Quantum Lighter, Classic Belt Lighter Case, leather. From Catalog 6, 1995. Zippo: $35-40; Military: $8-15; leather case: $5-10. *Courtesy of Odell Farley.*

1st Annual Camel Classic Golf Balls and Canister with a golfing Joe Camel on each one. On the right is the lid of the container. From Catalog 6, 1995. $10-12. *Courtesy of Odell Farley.*

Camel Pocket Pack, maple holder for ten regular-sized Camels. 3.75" x 4.5". From Catalog 6, 1995. $5-8. *Courtesy of Odell Farley.*

Camel Pool Rack Tin and a pack of Pool Player Playing Cards. From Catalog 6, 1995. $8-10. *Courtesy of Odell Farley.*

Camel Trench II Lighter. Catalog 6, 1995. $8-12. *Courtesy of Odell Farley.*

Hard Pack Tin-O-Coasters, set of four with different designs featuring Bustah, Floyd, Eddie, and Max. Cork bottom. From Catalog 6, 1995. $5-8. *Courtesy of Odell Farley.*

Smokin' Joe' Travel Mug, Camel Power. 12 oz. plastic mug made by Aladdin. Catalog 6, 1995. $7-10. *Courtesy of Odell Farley.*

140

Classic Camel Desert Bag, canvas, 18" x 23.5". From Catalog 6, 1995. $15-20. *Courtesy of Odell Farley.*

Hard Pack Denim shirt. From Catalog 6, 1995. $20-25. *Courtesy of Odell Farley.*

Joe's Denim Biker jacket. From Catalog 6, 1995. $35-50. *Courtesy of Odell Farley.*

Chapter 5:
The Camel Pack

Except for the last few years, when Camel began producing special collector packs the look of Camel packaging has hardly changed at all. Preproduction wrappers showed a rather silly looking rendition of the camel, but that was changed when Reynolds finagled the right to photograph Old Joe of the Barnum and Bailey Circus. Before the cigarette went into production a dignified image of Old Joe was firmly and permanently ensconced on the package.

The following are photographs of the packs Camel has used from its first production until the special limited edition Christmas pack it produced for Camel Filters and Camel Lights in 1995. All of them are from the collection of Odell Farley.

Note: *Cigarette packs range from current retail value to $35 per pack and $150 for a carton. The older or rarer the pack the greater the value. Any pack that is not currently in production is worth at least $5.*

Camels

  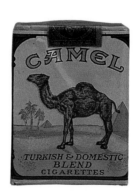

Three early packs. On the left is a paper and foil pack with no cellophane, c. 1917. Not the variegated sky; as time went on the sky grew more and more monotoned. In the middle is the paper and foil pack with cellophane, introduced 1931 to promote freshness; it does not yet have an opening tab. At the right is a pack from the war years, c. 1942-1946. It has cellophane but no foil because of the war. *Courtesy of Odell Farley.*

The oldest Camel carton, two piece box. c. 1913. *Courtesy of Odell Farley.*

Full World War II carton. One piece with fold in lid. As the war continued a switch was made from foil to paper. Very rare. *Courtesy of Odell Farley.*

Tin flat fifties, c. 1930s.

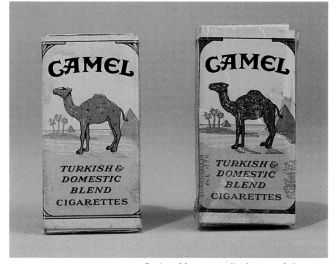

Packs of four, no cellophane or foil, c. 1920s. *Courtesy of Odell Farley.*

A flat fifties cardboard box. c. 1940s. *Courtesy of Odell Farley.*

Humidor tins of 100 and tin of 50. 1930s. *Courtesy of Odell Farley.*

In 1957 Reynolds attempted to modernize the pack. Consumers did not accept the change, and Reynolds soon switched back to the classic design. *$12-15. Courtesy of Odell Farley.*

Camel King-Size (85 mm) without filters were tested around 1965, but never made it to market. *Courtesy of Odell Farley.*

Chronological order after 1957 failure to the present day pack on the far right. Some dating of the packages can be made using the warning labels imposed by the FTC. They are: "Cigarette Smoking May Be Hazardous to Your Health" (1/1/66-10/31/70); "The Surgeon General Has Determined That Cigarette Smoking Is Dangerous to Your Health" (11/1/70-10/11/85); and "SURGEON GENERAL'S WARNING:" followed by one of four warnings..."Smoking Cause Lung Cancer, Heart Disease, Emphysema, and May Complicate Pregnancy"; "Quitting Smoking Now Greatly Reduces Serious Risks to Your Health"; "Smoking by Pregnant Women May Result in Fetal Injury, Premature Birth, and Low Birth Weight"; or "Cigarette Smoke Contains Carbon Monoxide" (10/12/85 to the present). *Courtesy of Odell Farley.*

King size Camel non-filter soft pack. *Courtesy of Odell Farley.*

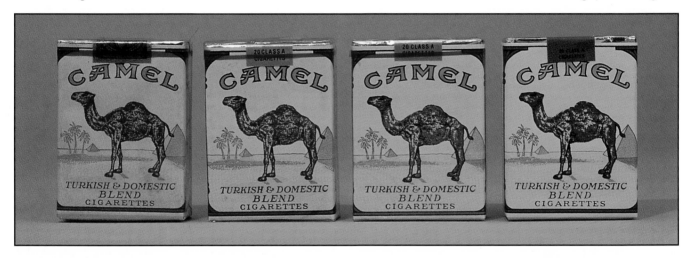

80th Anniversary carton, 1993. *Courtesy of Odell Farley.*

144

Camel Filters

This gold pack was one of three initial designs test marketed for the introduction of Camel Filters in 1965. The others were a brown, white and red version, and a filter version of the traditional pack. *Courtesy of Odell Farley.*

Actual pack used for Camel Filters, 85 mm, when introduced in August, 1966. *Courtesy of Odell Farley.*

When the gold design was eliminated, a box with three varieties of the brown and white design, and a traditional design was given to the executives to get their preference for the design choice. October, 1965. Box has the name of Mr. W.P. Hanes, written on it. Left to right: the winner; wood grain; linen finish; traditional.

First crushproof boxes of Camel Filters, in full pack, twelve pack, and four pack. The hard pack was introduced in October, 1975. *Courtesy of Odell Farley.*

Camel Filters carton, c. 1966. *Courtesy of Odell Farley.*

The same group showing the back side of the twelve pack of Camel filters. This was the first use of four-color back by Camels. *Courtesy of Odell Farley.*

Camel filter packs changed to the traditional design c. 1979. Here they are from then until 1988. Third one in is a complimentary twelve pack. *Courtesy of Odell Farley.*

Front and back of the 75th Anniversary Collector's pack, 1988. *Courtesy of Odell Farley.*

Collector's packs, featuring ten states, 1994. *Courtesy of Odell Farley.*

Current Camel filter pack with Genuine Taste slogan and Camel Cash, 1995. *Courtesy of Odell Farley.*

1995 limited edition Christmas pack sent to the best Camel Cash customers. 948,477 were sent to individuals, 12000 to the RJR gift store. $5-7. *Courtesy of Odell Farley.*

Camel filter menthol, 85 mm, August, 1966. They did not last for long, being officially discontinued in 1968 without every going national. Carton, full pack, and sample pack. *Courtesy of Odell Farley.*

Chronological order of Camel Filter hard packs, to the present day, and including the carton of Collector's packs. 1994. *Courtesy of Odell Farley.*

100s

Camel Talls, August, 1971. Introduced into test markets, this 100 mm filter cigarette was discontinued on January 18, 1972 without going national. *Courtesy of Odell Farley.*

Camel Filter 100's, essentially an updated version of Camel Talls, came onto the market in 1987. *Courtesy of Odell Farley.*

Camel Filter 99s, same as a 100, but in a crushproof box the cigarette had to be reduced one mm to allow for the cardboard, 1991. *Courtesy of Odell Farley.*

Lights

The first pack of Camel lights, 1979. *Courtesy of Odell Farley*

The first Camel Lights soft pack carton, 1979. *Courtesy of Odell Farley.*

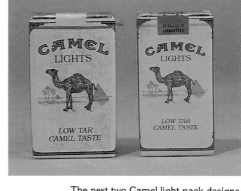

The next two Camel light pack designs, c. 1980 before 1988. *Courtesy of Odell Farley.*

Front and back of the 1988 Camel Lights 75th Anniversary Collector's pack. *Courtesy of Odell Farley.*

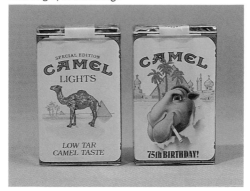

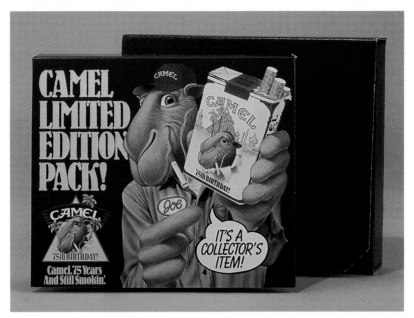

Presentation box of the 75th Anniversary
Collector's pack. *Courtesy of Odell
Farley.*

Collector's Camel Lights (soft packs)
featuring Joe's Place with Max, Eddie,
Floyd, Bustah, and Joe, 1993. *Courtesy
of Odell Farley.*

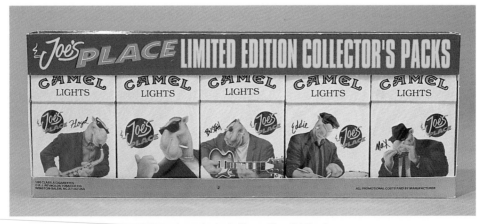

Collector's pack, Joe's Place, Max, Eddie, Joe,
Bustah, and Floyd. *Courtesy of Odell Farley.*

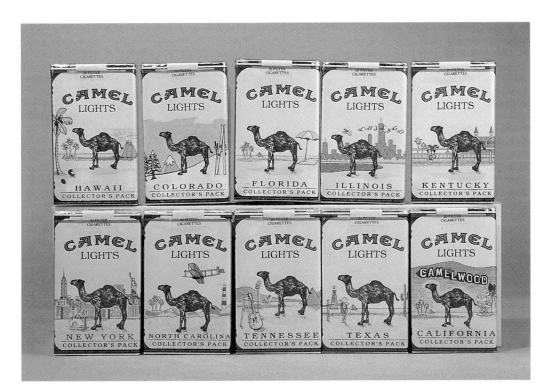

Collector's Camel
Lights (soft packs)
featuring ten states.
*Courtesy of Odell
Farley.*

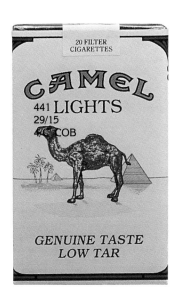

1995 Camel Light soft
pack. *Courtesy of
Odell Farley.*

The first Camel Lights Hard
Pack, pack of twelve. 1980
Courtesy of Odell Farley.

The back of the first Camel
Lights Hard Pack of twelve,
announcing its arrival. *Courtesy
of Odell Farley.*

Camel Lights, 1985-
1994. *Courtesy of
Odell Farley.*

Camel Lights Hard Pack Collector's series featuring Joe's Place with Max, Eddie, Floyd, Bustah, and Joe. *Courtesy of Odell Farley.*

1995 Camel Lights Hard Pack, Genuine Taste. *Courtesy of Odell Farley.*

The Camel Lights Hard Pack Collector's pack featuring ten states. 1994. *Courtesy of Odell Farley.*

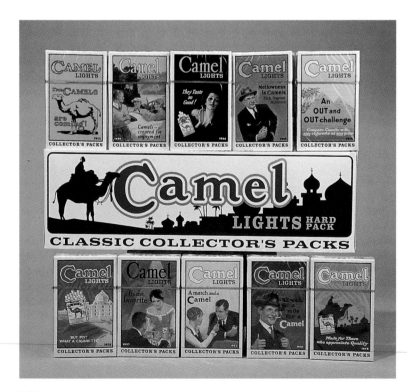

1995 Camel Lights Collector's pack, featuring miniature reproductions of classic Camel advertising. *Courtesy of Odell Farley.*

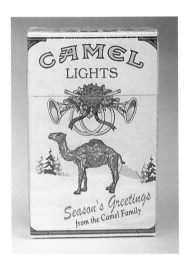

1995 limited edition Christmas pack sent to the best Camel Cash customers. 948,477 were sent to individuals, 12000 to the RJR gift store. $5-7. *Courtesy of Odell Farley.*

The box that contained the Christmas pack. $7-9 *Courtesy of Odell Farley.*

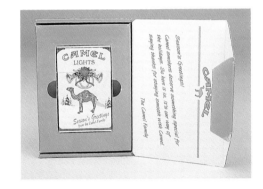

Camel Lights 100

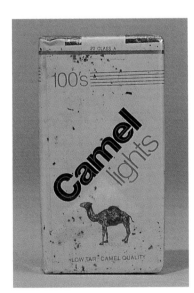

The first Camel Lights 100s soft pack. 1979. *Courtesy of Odell Farley.*

Redesigned after 1979, this is the front and back of the Camel Lights 100s soft pack, showing the new advertising. *Pack of 12 cigarettes. Courtesy of Odell Farley.*

Camel Lights 100s soft pack cigarettes, post-1979 to 1995. *Courtesy of Odell Farley.*

Camel Light 99s, same as a 100, but in a crushproof box, and now a length of 99 mm to allow room for the cardboard. *Courtesy of Odell Farley.*

Ultra Lights

Camel Ultra-Lights in the soft pack, left c. 1990, right 1995.

Camel Ultra-Light 100s, left 1990, right with convertible box, 1995. Convertible box was introduced in 1993. *Courtesy of Odell Farley.*

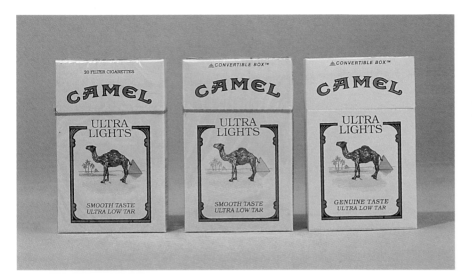

Camel Ultra-Lights in hard packs. The two on the right are in convertible boxes: can act as either a flip-top or soft pack, open top style. Left: c. 1990, center c. 1993, right: 1995. *Courtesy of Odell Farley.*

Wides

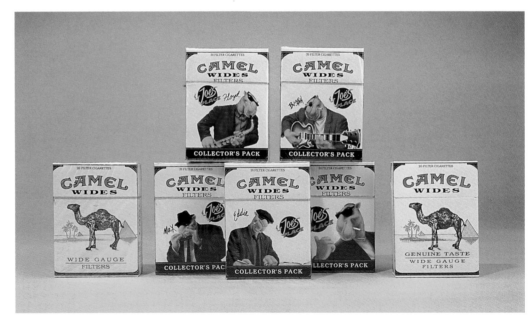

First box of Camel Wides on the left, 1992; the Hard Pack guys Collector's packs in the middle, 1993, and the 1995 pack on the right. *Courtesy of Odell Farley.*

Presentation box for Camel Wides, featuring Max and Ray, March 2, 1992. *Courtesy of Odell Farley.*

The ten states Collector's packs in the Camel Wide style, 1994. *Courtesy of Odell Farley.*

Camel Wide Lights. Left to right: The first pack, 1992, the Collector's ten states carton, 1994, and the current 1995 pack. *Courtesy of Odell Farley.*

Camel Special Lights, 85mm, soft pack. Left, 1993, right, 1995. *Courtesy of Odell Farley.*

Presentation box for Special Lights, April 5, 1993. *Courtesy of Odell Farley.*

Camel Special Lights in the box. Left, 1993; center, 1994; right, 1995. *Courtesy of Odell Farley.*

Camel Special Lights 100s in box. Left: 1993; right: 1995. *Courtesy of Odell Farley.*

World Variations

Foreign packs with minor differences. The one on the right is a twenty-five pack from Germany. *Courtesy of Odell Farley.*

Hard packs from various countries.
Courtesy of Odell Farley.

Arabian version of the Camel hard
pack, could not put the minarets on
the back, so the pyramid design was
repeated as shown here on the right.
Courtesy of Odell Farley.

70 mm Camels with filters made in
Mexico. *Courtesy of Odell Farley.*

Opposite page:
A variety of insert cards for packs of
Camels. Donated to soldiers during
World War II. *Courtesy of Odell
Farley.*

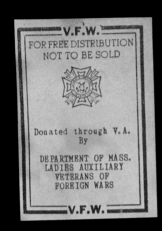

V.F.W.

FOR FREE DISTRIBUTION
NOT TO BE SOLD

Donated through V.A.
By

DEPARTMENT OF MASS.
LADIES AUXILIARY
VETERANS OF
FOREIGN WARS

V.F.W.

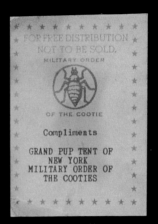

FOR FREE DISTRIBUTION
NOT TO BE SOLD.

MILITARY ORDER

OF THE COOTIE

Compliments

GRAND PUP TENT OF
NEW YORK
MILITARY ORDER OF
THE COOTIES

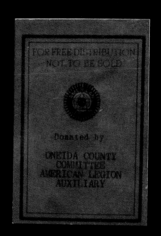

FOR FREE DISTRIBUTION
NOT TO BE SOLD

Donated by

ONEIDA COUNTY
COMMITTEE
AMERICAN LEGION
AUXILIARY

HAPPY FOURTH

COMPLIMENTS OF

SPECIAL SERVICES
RECEIVING STATION
CHARLESTON, S.C.

HOLIDAY GREETINGS

MILITARY

SEA TRANSPORTATION

SERVICE

ATLANTIC AREA

MADIGAN
N.C.O.
OPEN MESS

" HOLIDAY
GREETINGS"

COMPLIMENTS OF

Bimco

CORPORATION

1007 N. Liberty St.
WINSTON-SALEM, N. C.
Phone 4-2431
Wholesale
Plumbing & Heating
Supplies

OFFICER'S MESS
GREAT FALLS
AIR FORCE BASE
GREAT FALLS, MONT.

SEASON'S GREETING

Season's
Greetings

COMPLIMENTS OF

WESTERN HOTEL
CALLICOON, N. Y.
PHONE 45

MERRY CHRISTMAS
AND A
HAPPY NEW YEAR

WELCOME
ABOARD AIR FORCE
ONE

WELCOME ABOARD
MARINE ONE

WELCOME ABOARD
THE SEQUOIA

WELCOME ABOARD
PRESIDENTIAL
AIRCRAFT

WELCOME TO
CAMP DAVID

WELCOME
TO THE STAFF
MESS

Opposite page:
Insert cards for pack of cigarettes given to guests by the President of the United States. The gold card in the top center is from the 1940s or 1950s, while the others date to the Kennedy era. *Courtesy of Odell Farley.*

Two country insert cards. Left: Kuwait; right: Lebanon. Late 1960s. *Courtesy of Odell Farley.*

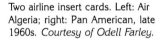

Two airline insert cards. Left: Air Algeria; right: Pan American, late 1960s. *Courtesy of Odell Farley.*

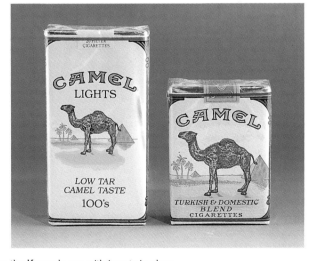

Unopened Presidential packs of Camels from the Kennedy era, with inserts in place.

Bibliography

Collins, Philip. *Smokerama: Classic Tobacco Accoutrements.* San Francisco: Chronicle Books, 1992.

Congdon-Martin, Douglas & Jerry Terranova: *Antique Cigar Cutters & Lighters.* Atglen, PA: Schiffer Publishing, 1996

Congdon-Martin, Douglas. *Tobacco Tins.* Atglen, PA: Schiffer Publishing, 1992.

_____. *America For Sale: A Collector's Guide to Antique Advertising.* Atglen, PA: Schiffer Publishing, 1992.

_____. *Camel Cigarette Collectibles: The Early Years: 1913-1963.* Atglen, PA: Schiffer Publishing, 1996.

Golden Leaves: R.J. Reynolds Tobacco Co. and the Art of Advertising. Winston-Salem: R.J. Reynolds Tobacco Co., 1986.

Les Aventures publicitaires d'un Dromadarie (Once Upon A Time A Camel...) Paris: Union des arts décoratifs, 1992.

Mullen, Chris. *Cigarette Pack Art.* New York: St. Martin's Press, 1979.

Petrone, Gerard S. *Tobacco Advertising: The Great Seduction.* Atglen, PA: Schiffer Publishing, 1996.

R.J. Reynolds Company, *Our 100th Anniversary: 1875-1975.* Winston-Salem: 1975.

Smith, Jane Webb. *Smoke Signals: Cigarettes, Advertising, and the American Way of Life.* Richmond: The Valentine Museum, 1990.

Tennant, Richard B. *The American Cigarette Industry: A Study in Economic Analysis and Public Policy.* New Haven: Yale University Press, 1950.

Tilley, Nannie M. *The R.J. Reynolds Tobacco Company.* Chapel Hill: University of North Carolina Press, 1985.

Wagner, Susan. *Cigarette Country: Tobacco in American History and Politics.* New York: Praeger Publishers, 1971.